Graduate Texts in Mathematics **146**

Graduate Texts in Mathematics

continued after index

Douglas S. Bridges

Computability

A Mathematical Sketchbook

With 29 Illustrations

Springer-Verlag
New York Berlin Heidelberg London Paris
Tokyo Hong Kong Barcelona Budapest

Douglas S. Bridges
Department of Mathematics
University of Waikato
Private Bag 3105
Hamilton, New Zealand

Editorial Board

Mathematics Subject Classifications (1991): 03Dxx

Library of Congress Cataloging-in-Publication Data
Bridges, D.S. (Douglas S.), 1945–
    Computability : a mathematical sketchbook / Douglas S. Bridges.
      p.   cm. — (Graduate texts in mathematics)
    Includes bibliographical references and index.
    ISBN 0-387-94174-6
    1. Computable functions. I. Title. II. Series.
QA9.59.B75   1994
511.3 – dc20   93-21313

Printed on acid-free paper.

Production managed by Hal Henglein; manufacturing supervised by Vincent Scelta.
Photocomposed pages prepared from the author's LaTeX file.
Printed and bound by R.R. Donnelley & Sons, Harrisonburg, VA.
Printed in the United States of America.

9 8 7 6 5 4 3 2 1

ISBN 0-387-94174-6 Springer-Verlag New York Berlin Heidelberg
ISBN 3-540-94174-6 Springer-Verlag Berlin Heidelberg New York

For Vivien, Iain, Hamish, and Catriona

'I can't believe that!' said Alice. 'Can't you?' the Queen said in a pitying tone. 'Try again: draw a long breath and shut your eyes.' Alice laughed. 'There's no use trying,' she said: 'One can't believe impossible things.' 'I daresay you haven't had much practice,' said the Queen.

LEWIS CARROLL, *Through the Looking Glass.*

Preface

My intention in writing this book is to provide mathematicians and mathematically literate computer scientists with a brief but rigorous introduction to a number of topics in the abstract theory of computation, otherwise known as *computability theory* or *recursion theory*. It develops major themes in computability, such as Rice's Theorem and the Recursion Theorem, and provides a systematic account of Blum's abstract complexity theory up to his famous Speed-up Theorem.

A relatively unusual aspect of the book is the material on computable real numbers and functions, in Chapter 4. Parts of this material are found in a number of books, but I know of no other at the senior/beginning graduate level that introduces elementary recursive analysis as a natural development of computability theory for functions from natural numbers to natural numbers.[1] This part of the book is definitely for mathematicians rather than computer scientists and has a prerequisite of a first course in elementary real analysis; it can be omitted, without rendering the subsequent chapters unintelligible, in a course including the more standard topics in computability theory found in Chapters 4-6.

I believe, against the trend towards weighty, all-embracing treatises (*vide* the typical modern calculus text), that many mathematicians would like to be able to purchase books that give them insight into unfamiliar branches of the subject in a relatively short compass and without requiring a major investment of time, effort, or money. Following that belief, I have had to exclude from this book many topics—such as detailed proofs of the equivalence of various mathematical models of computation, the theory of degrees of unsolvability, and polynomial and nonpolynomial complexity—whose absence will be deplored by at least some of the experts in the field. I hope that my readers will be inspired to pursue their study of recursion theory in such major works as [9, 24, 28, 29].

A number of excellent texts on computability theory are primarily aimed at computer scientists rather than mathematicians, and so do not always maintain the level of rigour that would be expected in a modern text on, say, abstract algebra. I have tried to maintain that higher level of rigour

---

[1] Some of the work in this book—notably, Proposition (4.28) and the application of the Recursion Theorem preceding Exercises (5.14)—appears to be original.

throughout, even at the risk of deflecting the interest of mathematically insecure computer scientists.

Ideally, all mathematics and computer science majors should be exposed to at least some of the material found in this book. It horrifies me that in some universities such majors can still graduate ignorant of the theoretical limitations of the computer, as expressed, for example, by the undecidability of the halting problem (Theorem (4.2)). A short course on computability, accessible even to students below junior level, would comprise Chapters 1-3 and the material in Chapter 4 up to Exercises (4.7). A longer course for more advanced undergraduates would also include Rice's theorem and the Recursion Theorem, from Chapter 5, and at least parts of Chapter 6. The entire book, including the difficult material on recursive analysis from Chapter 4, would be suitable for a course for bright seniors or beginning graduate students.

I have tried to make the book suitable for self-study. To this end, it includes solutions for most of the exercises. Those exercises for which no solutions are given have been marked with the asterisk (*); of varying levels of difficulty, they provide the instructor with material for homework and tests. *The exercises form an integral part of the book* and are not just there for the student's practice; many of them develop material that is used in later proofs, which is another reason for my inclusion of solutions.

My interest in constructive mathematics [5] leads me to comment here on the logic of computability theory. This is *classical logic,* the logic used by almost all mathematicians in their daily work. However, the use of classical logic has some perhaps undesirable consequences. Consider the following definition of a function f on the set $\mathbf{N}$ of natural numbers: for all n, $f(n)$ equals 1 if the Continuum Hypothesis is true, and equals 0 if the Continuum Hypothesis is false.[2] Since 'most mathematicians are formalists on weekdays and Platonists on Sundays', at least on Sundays most of us would accept this as a good definition of a function f. According to classical logic, f is computable because there exists an algorithm that computes it; that algorithm is either the one which, applied to any natural number n, outputs 1, or else the one which, applied to any natural number n, outputs 0. But the Continuum Hypothesis is independent of the axioms of ZFC (Zermelo-Fraenkel set theory plus the axiom of choice), the standard framework of mathematics, so we will never be able to tell, using ZFC alone, which of the two algorithms actually is the one that computes f.

It appears from this example, eccentric though it may be, that the standard theory of computation does not exactly match computational practice,

---

[2]The **Continuum Hypothesis** (CH) says that the smallest cardinal number greater than $\aleph_0$, the cardinality of $\mathbf{N}$, is $2^{\aleph_0}$, the cardinality of the set of all subsets of $\mathbf{N}$. The work of Cohen [13] and Gödel [17] shows that neither CH nor its negation can be proved within Zermelo-Fraenkel set theory plus the axiom of choice; see also [3], pages 420-428.

in which we would expect to pin down the algorithms that we use. A facetious question may reinforce my point: what would happen to an employee who, in response to a request that he write software to perform a certain computation, presented his boss with two programs and the information that, although one of those programs performed the required computation, nobody could ever tell which one?

With classical logic there seems to be no way to distinguish between functions that are computed by programs which we can pin down and those that are computable but for which there is no hope of our telling which of a range of programs actually performs the desired computation. To handle this problem successfully, we need a different logic, one capable of distinguishing between *existence in principle* and *existence in practice*. For example, with constructive (intuitionistic) logic the problem disappears,[3] since f is then not properly defined: it is only properly defined if we can decide the truth or falsehood of the Continuum Hypothesis (which we cannot) and therefore which of the two possible algorithms computes f.

Having said this, let me stress that, despite the inability of classical logic to make certain distinctions of the type I have just dealt with, *I have followed standard practice and used classical logic throughout this book.*

Not only the logic but also most of the material that I have chosen is standard, although some of the exercises and examples are new. I have drawn on a number of books, including [34] for the treatment of Turing machines in Chapter 1; [20] for the first parts of Chapters 4 and 5; and [9, 14, 29] for parts of Chapter 7.

The origins of my book lie in courses I gave at the University of Buckingham (England), New Mexico State University (USA), and the University of Waikato (New Zealand). I am grateful to the students in those classes for the patience with which they received various slowly improving draft versions.[4] Special thanks are due to Fred Richman for many illuminating conversations about recursion theory; to Paul Halmos for his advice and encouragement; and to Cris Calude, Nick Dudley Ward, Graham French, Hazel Locke, and Steve Merrin, all of whom have read versions of the text and made many helpful corrections and suggestions. As always, it is my wife and children who suffered most as the prolonged birth of this work took so much of my care and attention; I present the book to them with love and gratitude.

May 1993   *Douglas S. Bridges*

---

[3]For a development of computability theory using intuitionistic logic see Chapter 3 of [8].

[4]The first drafts of this book were prepared using the T^3 *Scientific Word Processing System*. The final version was produced by converting the drafts to T_EX and then using *Scientific Word*. T^3 and *Scientific Word* are both products of TCI Software Research, Inc. The diagrams were drawn with *Aldus Freehand* v. 3.1 (©Aldus Corporation).

Contents

Preliminaries

Throughout this book we assume familiarity with the standard notations and basic results of informal set theory, as found in [18]. We use the following notation for sets of numbers.

The set of natural numbers: $\mathbf{N} \equiv \{0, 1, 2, \ldots\}$.

The set of rational numbers: $\mathbf{Q} \equiv \{\pm m/n : m, n \in \mathbf{N}, \ n \neq 0\}$.

The set of real numbers: $\mathbf{R}$.

For $n \geq 1$ we write X^n for the n−fold Cartesian product $X \times X \times \cdots \times X$ (n factors) of X, and P_i^n for the ith projection of X^n—that is, the mapping from X^n onto X defined by[1]

$$P_i^n(x_1, \ldots, x_n) \equiv x_i.$$

We denote by $(x_n)_{n=0}^{\infty}$, or $(x_0, x_1, \ldots)$, or even just (x_n), the sequence whose terms are indexed by $\mathbf{N}$ and whose n^{th} term is x_n.

We shall be particularly interested in what happens when a computer is programmed to compute natural number outputs from inputs in $\mathbf{N}^n$. Since the execution of a program may fail to terminate when the machine is run with certain inputs—for example, a program for computing the reciprocal of a natural number will not normally output a natural number if it is run with the input 0—we are forced to deal with functions that are defined on subsets of $\mathbf{N}^n$ and not necessarily on the entire set $\mathbf{N}^n$. This leads us to the notion of a **partial function** φ from a set A to a set B : that is, a function φ whose domain is a subset of A and which takes values in B; the domain of φ may be empty and is usually not the entire set A. We refer to such a function φ as the partial function $\varphi : A \to B$; we write domain(φ) for its domain, and range(φ) for its range. We also say that $\varphi(x)$ is **defined** if $x \in$ domain(φ), and that $\varphi(x)$ is **undefined** if $x \in A$ and $x \notin$ domain(φ). A partial function from A to B whose domain is the entire set A is called, oxymoronically, a **total partial function** from A to B.

There is an unwritten convention (not followed by all authors) that uses Greek letters to denote partial functions and Roman letters to denote total ones. We shall usually follow that convention, although some partial func-

---

[1] The symbol $\equiv$ means *is defined as* or *is identical to*.

tions that are not initially known to be total and are therefore denoted by Greek letters will eventually turn out to be total.

We often give explicit definitions of partial functions, of the following form[2]:

$$\varphi(n) \;\; = \;\; \sqrt{n} \qquad \text{if } n \text{ is a perfect square,}$$
$$= \;\; \text{undefined} \quad \text{otherwise.}$$

In this example, φ is a partial function from $\mathbf{N}$ to $\mathbf{N}$, and

$$\text{domain}(\varphi) = \{n \in \mathbf{N} : n \text{ is a perfect square}\}.$$

We also describe φ as the partial function $n \mapsto \sqrt{n}$ from $\mathbf{N}$ to $\mathbf{N}$. We use the arrow $\rightarrow$ as in 'the partial function $\varphi : A \rightarrow B$', and the barred arrow $\mapsto$ as in 'the partial function $x \mapsto \sin x$ on $\mathbf{R}$'. If, for example, Φ is a partial function from $\mathbf{N}^2$ to $\mathbf{N}$, then for each $m \in \mathbf{N}$ we also denote the partial function $n \mapsto \Phi(m, n)$ on $\mathbf{N}$ by $\Phi(m, \cdot)$.

Partial functions can be operated on in the obvious ways. For example, if φ, ψ are partial functions from $\mathbf{N}$ to $\mathbf{N}$, then their **sum**, **product**, and **composite** are defined respectively as follows:

$$(\varphi + \psi)(n) \;\; = \;\; \varphi(n) + \psi(n) \quad \text{if } \varphi(n), \psi(n) \text{ are both defined,}$$
$$= \;\; \text{undefined} \qquad \text{otherwise;}$$

$$(\varphi \cdot \psi)(n) \;\; = \;\; \varphi(n) \cdot \psi(n) \quad \text{if } \varphi(n), \psi(n) \text{ are both defined,}$$
$$= \;\; \text{undefined} \qquad \text{otherwise;}$$

$$\varphi \circ \psi(n) \;\; = \;\; \varphi(\psi(n)) \qquad \text{if } \psi(n) \text{ is defined and belongs}$$
$$\text{to domain}(\varphi),$$
$$= \;\; \text{undefined} \qquad \text{otherwise;}$$

In general, if a partial function $\Phi : \mathbf{N}^m \rightarrow \mathbf{N}^n$ is defined in terms of already constructed partial functions $\varphi_i : \mathbf{N}^m \rightarrow \mathbf{N}$ $(1 \le i \le j)$ and $\Psi : \mathbf{N}^j \rightarrow \mathbf{N}^n$ by an equation of the type

$$\Phi(k_1, \ldots, k_m) = \Psi(\varphi_1(k_1, \ldots, k_m), \ldots, \varphi_j(k_1, \ldots, k_m)),$$

it is assumed that the left hand side is defined if and only if the right hand side is defined; thus

$$\text{domain}(\Phi) \;\; = \;\; \{u \in \mathbf{N}^m : u \in \bigcap_{i=1}^{j} \text{domain}(\varphi_i) \text{ and}$$
$$(\varphi_1(u), \ldots, \varphi_j(u)) \in \text{domain}(\Psi)\}.$$

Likewise, when we write

$$\varphi(n) \le k,$$

---

[2]Here and throughout the book, we use $\sqrt{n}$ to denote the nonnegative square root of a nonnegative real number.

where $\varphi : \mathbf{N} \to \mathbf{N}$ is a partial function, we imply that $\varphi(n)$ *is defined* and less than or equal to k.

For computational purposes a natural number n is usually represented by a string of symbols drawn from some suitable set. For example, 5 may be represented by the string $aaaaa$ whose symbols are drawn from the singleton set $\{a\}$, by the binary string 101, by the single decimal digit 5, and so on. Strings appear so frequently in the early chapters of our book that it is a good idea to give a formal definition of them here.

By a **string of length** n over the set X we mean an element $(x_1, \ldots, x_n)$ of the $n-$fold Cartesian product $X^n \equiv X \times \cdots \times X$ $(n$ factors$)$; the elements $x_1, \ldots, x_n$ are called the **terms** of the string, x_k being the kth term. When we consider $(x_1, \ldots, x_n)$ as a string over X, we usually omit the parentheses and commas, and simply write $x_1 \ldots x_n$. We assume that there is a unique **empty string** Λ of length 0 over X; informally, Λ is the unique string with no terms over X. We denote by X^* the set of strings over X, and by $|u|$ the length of the string $u \in X^*$.

Strings u, v over X can be combined to form a string $u \cdot v$, usually written uv, by the operation of **concatenation.** Informally, this involves writing one string next to another. The following is a formal inductive definition: for all strings u, v over X, and all elements x of X,

$$\begin{aligned} \Lambda \cdot u &\equiv u, \\ (xu) \cdot v &\equiv x(u \cdot v). \end{aligned}$$

It is a simple exercise in induction to show that concatenation has the properties you would expect it to have. For example, the length of uv is the sum of the lengths of u and v; $u(vw) = (uv)w$ (so we write either side as uvw); and

$$\Lambda u = u = u \Lambda.$$

In the context of computability and formal language theory a nonempty finite set X is often called an **alphabet**, and a subset of X^* a **language** over X. (The set of words defined in the *Oxford English Dictionary* is a language over the alphabet $\{a, b, c, \ldots, z\}$; British readers might argue that this is *the* English language!) The following are useful constructions with languages A, B over a finite alphabet X :

- The **concatenation** of A and B :

$$A \cdot B \equiv \{uv : u \in A, \, v \in B\}.$$

- The **iterate** (or **Kleene star**) of A :

$$A^* \equiv \{u_1 u_2 \ldots u_n : n \geq 0, \, \forall k \, (u_k \in A)\}.$$

In this context the union of A and B is often written $A + B$, rather than $A \cup B$.

For example, if $X = \{0, 1, 2\}$, $A = \{0, 1\}^*$, and $B = \{2\}$, then $A \cdot B$ consists of the string 2, together with all strings of the form $x_1 \ldots x_n 2$ with $x_i \in \{0, 1\}$ for $1 \leq i \leq n$; $A^* = A$, and B^* consists of all finite (possibly empty) strings with each term equal to 2; and $A + B$ consists of all binary strings together with the single string 2.

There are common shorthand notations which avoid cumbersome expressions for combinations of languages. For example, we write AB instead of $A \cdot B$,

$$010^*1^*0 \text{ instead of } \{01\} \cdot \{0\}^* \cdot \{1\}^* \cdot \{0\},$$

and

$$abb(ab)^* + (a + b)^*ba^* \text{ instead of } \{abb\} \cdot \{ab\}^* + \{a, b\}^* \cdot \{b\} \cdot \{a\}^*.$$

1

What Is a Turing Machine?

A Turing machine is ... the ultimate personal computer, since only pencil and paper are needed ... at the same time, it is as powerful as any real machine. ([34], p. 280)

We begin our study of computability by describing one of the earliest mathematical models of computation, one for which the underlying informal picture is especially easy to understand—the Turing machine.

In that picture (see Figure 1), a Turing machine consists of an infinite tape, and a read/write head connected to a control mechanism. The tape is divided into infinitely many cells, each of which contains a symbol from an alphabet called the tape alphabet; this alphabet includes the special symbol **B** to signify that a cell is blank (empty). The cells are scanned, one at a time, by the read/write head, which can move in both directions as long as it does not move off the tape (which would happen if, for example, the tape was bounded on the left and the read/write head moved left from the leftmost cell). At any given time, the machine (or, more properly, its control mechanism) will be in one of a finite number of possible states. The behaviour of the read/write head, and the change, if any, of the machine's state, are governed by the present state of the machine and by the symbol in the cell under scan.

The machine operates on words over an input alphabet which is a subset of the tape alphabet. The symbols forming such a word are written, in order, in consecutive cells from the left of the tape. When the machine enters a state, the read/write head reads the symbol in the cell against which it rests, and writes in that cell a symbol from the tape alphabet; it then moves one cell to the left, or one cell to the right, or not at all; after that, the machine enters its next state.

In this model there is no direct counterpart to the memory registers of a computer. However, information is stored in the sequence of states through which the machine passes. For example, if we want a Turing machine to transfer the content of a certain cell to the adjacent cell on the right, we "memorise" the symbol s read from the first cell by passing to a different state for each possible choice of s.

We now give a formal definition of some of these notions. Let X, Y be finite alphabets with $X \subset Y$, and **B** a distinguished **blank** element of

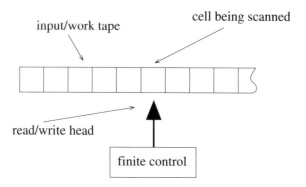

FIGURE 1. A Turing machine.

$Y\backslash X$. A **Turing machine** with **tape alphabet** Y and **input alphabet** X is a quadruple $\mathcal{M} \equiv (\mathcal{Q}, \delta, q_0, q_F)$ consisting of

- a finite set $\mathcal{Q}$ of **states**,

- a partial function $\delta : \mathcal{Q} \times Y \to \mathcal{Q} \times Y \times \{L, R, \Lambda\}$—the **state transition function**,

- a **start state** $q_0 \in \mathcal{Q}$, and

- a **halt state** $q_F \in \mathcal{Q}$,

where $\delta(q_F, y)$ is undefined for all y in Y.[1] We interpret the symbols L, R, and Λ as *left move, right move,* and *no move,* respectively.

We shall discuss examples of Turing machines later in the chapter. Our next task is to clarify our informal picture of the behaviour of a Turing machine.

In order to start a computation, the symbols of the input word

$$w \equiv x_1 \dots x_N \in X^*$$

must be written in the leftmost N cells of the tape, and $\mathcal{M}$ must be in the state q_0, with the read/write head against the leftmost cell. If $\mathcal{M}$ reads the symbol y in the state q, it computes $(q', y', D) = \delta(q, y)$, provided this quantity is defined. It then writes y'; moves left if $D = L$, right if $D = R$, not at all if $D = \Lambda$; and passes to the state q'. If $\mathcal{M}$ reaches the state

---

[1]Strictly speaking, we have defined here a **deterministic Turing machine**. This should be contrasted with a nondeterministic one, in which there is a choice of several actions when the machine reads a given symbol in a given state. Since we shall not be concerned with nondeterministic Turing machines, we shall use the shorter phrase *Turing machine*, rather than *deterministic Turing machine*, throughout this book.

q_F, its activity stops and the final output of its computation is read from the symbols remaining on the tape. (Actually, we need to be more careful about characterising the moves, halting behaviour, and outputs of $\mathcal{M}$; we will return to this matter shortly.)

Suppose that at a given instant our Turing machine $\mathcal{M}$ is in the state q; that the symbols in the cells on the left of the read/write head form the string $u \in Y^*$; that the terms of a string $v \in Y^*$ lie in the cells at, and to the right of, the read/write head; and that all cells to the right of v are blank. Thus the leftmost cells of the tape contain the string uv, and all cells to the right of this are blank. Then the instantaneous configuration of the machine is fully described by the triple (u, q, v), and the state transitions of $\mathcal{M}$ can be described by the sequence of triples giving the configurations of $\mathcal{M}$ at successive instants of the computation.

In order to formalise these ideas, we introduce two intuitively computable functions from Y^* to Y :

$$
\begin{aligned}
\mathbf{lend}(v) &= \mathbf{B} \quad \text{if } v = \Lambda, \\
&= c \quad \text{if } v = cw, c \in Y, \text{ and } w \in Y^*,
\end{aligned}
$$

$$
\begin{aligned}
\mathbf{rend}(v) &= \mathbf{B} \quad \text{if } v = \Lambda, \\
&= c \quad \text{if } v = wc, \ c \in Y, \text{ and } w \in Y^*.
\end{aligned}
$$

(Thus if v is a nonempty string over Y, then $\mathbf{lend}(v)$ is the leftmost symbol, and $\mathbf{rend}(v)$ is the rightmost symbol, of v.) Next, we define a **configuration** of $\mathcal{M}$ to be a triple (u, q, v), where $u \in Y^*$, either $v = \Lambda$ or $v \in Y^*(Y \backslash \{\mathbf{B}\})$, and $q \in \mathcal{Q}$.[2] We say that the configuration (u', q', v') is **reached in one step** from (u, q, v) if

$$
\delta(q, \mathbf{lend}(v)) = (q', b, D) \in \mathcal{Q} \times Y \times \{L, R, \Lambda\}
$$

is defined (so, in particular, $q \neq q_F$), and if the following conditions obtain.

(i) If $D = L$, then $u = u'\mathbf{rend}(u)$ and

$$
\begin{aligned}
v' &= \Lambda && \text{if } b = \mathbf{B}, \mathbf{rend}(u) = \mathbf{B}, \text{ and} \\
& && \text{either } v = \Lambda \text{ or } v = \mathbf{lend}(v), \\
&= \mathbf{rend}(u) && \text{if } b = \mathbf{B}, \mathbf{rend}(u) \neq \mathbf{B}, \text{ and} \\
& && \text{either } v = \Lambda \text{ or } v = \mathbf{lend}(v), \\
&= \mathbf{rend}(u)bw && \text{if } w \in Y^*, v = \mathbf{lend}(v)w, \text{ and} \\
& && \text{either } b \neq \mathbf{B} \text{ or } w \neq \Lambda.
\end{aligned}
$$

---

[2]As it stands, this definition does not completely capture our intuitive conception of a configuration, since it does not preclude the possibility of nonblank symbols lying to the right of the string v on the tape. However (see Exercise (1.2.2)), this situation does not arise when the configuration (u, q, v) is part of a computation, according to the strict notion of computation that we shall introduce shortly.

(ii) If $D = \Lambda$, then $u' = u$ and

$$
\begin{aligned}
v' &= \Lambda && \text{if } v = \Lambda \text{ and } b = \mathbf{B}, \\
&= b && \text{if } v = \Lambda \text{ and } b \neq \mathbf{B}, \\
&= bw && \text{if } v \neq \Lambda \text{ and } v = \mathbf{lend}(v)w, \text{ where } w \in Y^*.
\end{aligned}
$$

(iii) If $D = R$, then $u' = ub$ and

$$
\begin{aligned}
v' &= \Lambda && \text{if } v = \Lambda, \\
&= w && \text{if } v \neq \Lambda \text{ and } v = \mathbf{lend}(v)w, \text{ where } w \in Y^*.
\end{aligned}
$$

We then write

$$(u, q, v) \vdash (u', q', v').$$

Figures 2 and 3 illustrate some of the cases of this rather complicated definition; in each case, $v \neq \Lambda$, $a = \mathbf{rend}(u)$, $c = \mathbf{lend}(v)$, and $b \neq \mathbf{B}$.

(1.1) Exercise

* Draw diagrams to illustrate the remaining cases of the definition of *reached in one step*.

For configurations C, C' and a positive integer i, we define the relation $\vdash^i$ inductively as follows: $C \vdash^i C'$ if

either $i = 1$ and $C \vdash C'$,

or $i > 1$ and there exists C'' such that $C \vdash^{i-1} C''$ and $C'' \vdash C'$.

If $C \vdash^i C'$, then C' is **reached in i steps** from C.

We say that a configuration (u, q, v) of $\mathcal{M}$ is **admissible** if

either $u \neq \Lambda$,

or $u = \Lambda$ and $P_3^3(\delta(q, \mathbf{lend}(v))) \neq L$;

otherwise, we say that the configuration is **inadmissible**. A **computation** by $\mathcal{M}$ is a finite sequence $(C_0, C_1, \ldots, C_n)$ of admissible configurations such that

$C_0 = (\Lambda, q_0, v)$ for some $v \in X^*$,

$C_i \vdash C_{i+1}$ for each i,

and C_n is of the form (Λ, q_F, v') for some $v' \in X^*$.

We then call v the **input**, v' the **output**, C_0 the **initial configuration**, and C_n the **final configuration** of the computation; and we say that $\mathcal{M}$ **completes the computation** $(C_0, C_1, \ldots, C_n)$ on the input v.

In allowing only admissible configurations in the definition of *computation*, we have in mind the model where the Turing machine has its tape

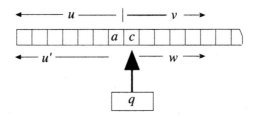

becomes

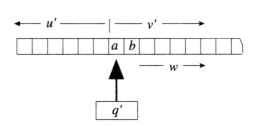

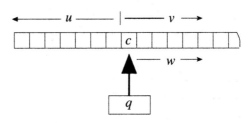

becomes

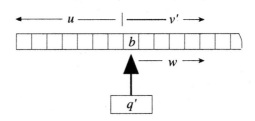

FIGURE 2. Two cases of passage from one configuration to another when $v \neq \Lambda, a = \mathbf{rend}(u), c = \mathbf{lend}(v)$, and $b \neq \mathbf{B}$.

$D=R$:

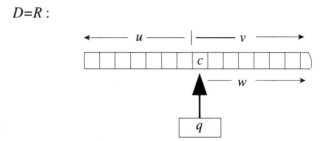

becomes

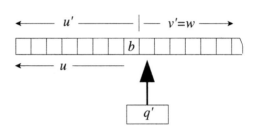

FIGURE 3. Another case of passage from one configuration to another when $v \neq \Lambda, a = \mathbf{rend}(u), c = \mathbf{lend}(v)$, and $b \neq \mathbf{B}$.

bounded on the left. In that model the input is written in the leftmost cells of the tape, and it is impossible to move left of the leftmost cell. The restriction to such Turing machines is not as drastic as it may seem, for it can be shown that to each Turing machine with tape extending infinitely in both directions there corresponds a Turing machine with tape bounded on the left that performs the same computation (usually using a different algorithm); see Section 6.4 of [34].

If $v \in X^*$ and there does not exist a computation by $\mathcal{M}$ with input v, we say that $\mathcal{M}$ **fails to complete a computation** on the input v. To see how such a failure can happen, let $C_0 \equiv (\Lambda, q_0, v)$, with $v \neq \Lambda$. If $P_3^3(\delta(q_0, \mathbf{lend}(v))) = L$, then C_0 is inadmissible; otherwise, there is a (possibly finite) sequence $C_0, C_1, \ldots$ of configurations such that $C_i \vdash C_{i+1}$ for each i. Either that sequence is infinite, in which case none of the states $P_1^3(C_i)$ is q_F; or the sequence is finite, with last term $C_n \equiv (u_n, q_n, v_n)$, say. In that event we have the following three possibilities: C_i is inadmissible for some i; $q_n \neq q_F$ and $\delta(q_n, \mathbf{lend}(v_n))$ is undefined; $q_n = q_F$. Only in the last case can $\mathcal{M}$ complete a computation—namely, $(C_0, C_1, \ldots, C_n)$—on the input v; even then, it only does so if $u_n = \Lambda$.

In general, if $v \in X^*$ and there is a finite sequence $(C_0, C_1, \ldots, C_n)$ of admissible configurations such that

$$C_0 = (\Lambda, q_0, v),$$
$$C_i \vdash C_{i+1} \text{ for each } i,$$
$$\text{and } C_n \text{ is of the form } (u', q_F, v') \text{ for some } u', v' \in Y^*,$$

we say that $\mathcal{M}$ **halts on the input** v. (Recall that q_F is the halt state of $\mathcal{M}$.) If C_n is of the form (Λ, q_F, v') for some v', we say that $\mathcal{M}$ **halts with the read/write head on the left**, or that $\mathcal{M}$ **parks the read/write head**; in which case, if also $v' \in X^*$, then $\mathcal{M}$ completes the computation $(C_0, C_1, \ldots, C_n)$ on the input v.

Many authors would not make the parking of the read/write head necessary for the completion of a computation: they would consider a computation $(C_0, C_1, \ldots, C_n)$ to be completed if C_n has the form (u', q_F, v') for some strings u', v' in X^*. We prefer to require the parking of the read/write head as this makes it easier to perform certain tasks such as the joining together of Turing machine modules[3] in the construction of a large Turing machine.

(1.2) Exercises

.1   Consider a Turing machine $\mathcal{M}$ with start state q_0 and halt state q_F. Suppose that from the initial configuration (Λ, q_0, v), where v is a string over the input alphabet X of $\mathcal{M}$, $\mathcal{M}$ follows a sequence of state transitions that eventually leave it in its halt state with the read/write head on the left, and with a string $v' \in X^*$, followed by a blank, in the leftmost cells of the tape. Can we decide whether $\mathcal{M}$ has completed a computation with output v'? In other words, can we determine whether there are nonblank symbols in cells of the tape to the right of v'?

.2* Prove that if the configuration $C' \equiv (u', q', v')$ is reached in one step from the configuration $C \equiv (u, q, v)$, and if, when the configuration of $\mathcal{M}$ is C, all cells to the right of the string uv are blank, then, after the transition from C to C', all cells to the right of $u'v'$ are blank.

The time has come to give some examples to clarify the many complicated definitions we have introduced above. To begin with, consider the Turing machine $\mathcal{M} \equiv (\mathcal{Q}, \delta, q_0, q_F)$ where

$$\mathcal{Q} = \{q_0, q_1, q_2, q_3, q_F\}, \ X = \{0, 1\}, \ Y = \{0, 1, \mathbf{B}\},$$

---

[3]When we refer to a *Turing machine module*, we have in mind the Turing machine equivalent of a procedure or subroutine in a programming language.

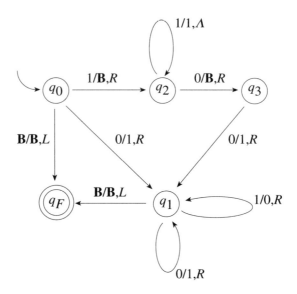

FIGURE 4. The state diagram of the Turing machine $\mathcal{M}$.

and δ is given by the following **state transition table**, in which, for example, the entry at the intersection of row q_1 and column **1** is $\delta(q_1, 1)$:

	0	**1**	**B**
q_0	$(q_1, 1, R)$	$(q_2, \mathbf{B}, R)$	$(q_F, \mathbf{B}, L)$
q_1	$(q_1, 1, R)$	$(q_1, 0, R)$	$(q_F, \mathbf{B}, L)$
q_2	$(q_3, \mathbf{B}, R)$	$(q_2, 1, \Lambda)$	undefined
q_3	$(q_1, 1, R)$	undefined	undefined
q_F	undefined	undefined	undefined

A more perspicuous representation is given by a directed graph known as the **state diagram** of $\mathcal{M}$; see Figure 4. In such a diagram the encircled nodes represent the states of the Turing machine. The initial state is distinguished as the one at the head of a curved arrow with no state at its tail, and the halt state by double encircling. An arrow bearing the label

$$y/y', D$$

and joining a state q to a state q' indicates that $\delta(q, y) = (q', y', D)$.

　　Now consider the behaviour of $\mathcal{M}$ when given the input 0011. The informal picture is given in Figure 5. A more formal description of the behaviour of $\mathcal{M}$ is given by the configuration sequence

$$(\Lambda, q_0, 0011) \quad \vdash \quad (1, q_1, 011)$$
$$\vdash \quad (11, q_1, 11)$$
$$\vdash \quad (110, q_1, 1)$$
$$\vdash \quad (1100, q_1, \Lambda)$$
$$\vdash \quad (110, q_F, 0).$$

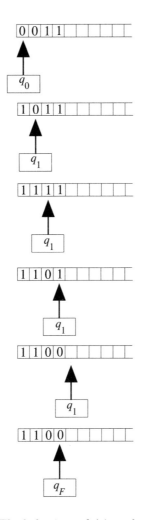

FIGURE 5. The behaviour of $\mathcal{M}$ on the input 0011.

Next, consider what happens when the input of $\mathcal{M}$ is 11; the informal picture is given in Figure 6.

Thus the machine *loops*, reading 1 in state q_2, writing 1, and remaining

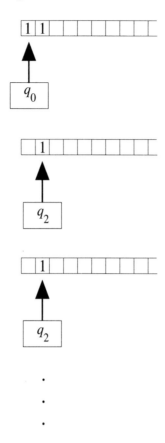

FIGURE 6. The behaviour of $\mathcal{M}$ on the input 0011.

in state q_2. The corresponding configuration sequence is

$$
\begin{aligned}
(\Lambda, q_0, 11) \quad &\vdash \quad (\mathbf{B}, q_2, 1) \\
&\vdash \quad (\mathbf{B}, q_2, 1) \\
&\vdash \quad (\mathbf{B}, q_2, 1) \\
&\vdash \quad \cdots .
\end{aligned}
$$

Finally, note that if $\mathcal{M}$ is given the empty input Λ, then the initial configuration (Λ, q_0, Λ) is inadmissible, since $P_3^3(\delta(q_0, \mathbf{lend}(\Lambda))) = L$. So $\mathcal{M}$ fails to perform a computation on the input Λ.

For our second example, we design a Turing machine $\mathcal{T}$, with input alphabet $\{0, 1\}$, that removes the leftmost symbol of the input word and shifts the resulting word one space to the left on the input tape. Here is an informal, high-level description of $\mathcal{T}$. It has states $q_0, q_1, q_2, q_3, q_4, q_F$, where q_0 is the start state and q_F the halt state. Assume that the symbols of a binary input string $w \equiv x_1 \ldots x_N$ are written in the leftmost N cells of the tape, and that T is in the state q_0, with the read/write head scanning the leftmost cell. If $N = 0$, so that $w = \Lambda$ and the symbol in the leftmost cell is **B**, T

> writes **B**,
> does not move, and
> enters the halt state q_F.

If $N \geq 1$ and therefore $w \neq \Lambda$, T first

> reads x_1,
> writes **B**,
> moves right, and
> enters the state q_1.

Now assume that T is in the state q_1, and that the cell on the left of the one scanned by the read/write head contains **B**. If the symbol scanned is an element x of $\{0, 1\}$, T

> writes **B**,
> moves left, and
> enters a state that depends on x (and thereby "memorises" x).

T then reads **B**,

> writes x,
> moves two cells right, and
> enters the state q_1.

If, on the other hand, T is in the state q_1, and the read/write head is scanning a cell that contains **B**, T

> writes **B**,
> does not move, and
> enters the halt state q_F.

Note that if the input string contains more than one term, T will not park the read/write head before entering its halt state. (You are invited to remedy this defect.)

The state transition table for T is given below; the state diagram is found in Figure 7.

	0	1	B
q_0	$(q_1, \mathbf{B}, R)$	$(q_1, \mathbf{B}, R)$	$(q_F, \mathbf{B}, \Lambda)$
q_1	$(q_2, \mathbf{B}, L)$	$(q_3, \mathbf{B}, L)$	$(q_F, \mathbf{B}, \Lambda)$
q_2	undefined	undefined	$(q_4, 0, R)$
q_3	undefined	undefined	$(q_4, 1, R)$
q_4	undefined	undefined	$(q_1, \mathbf{B}, R)$
q_F	undefined	undefined	undefined

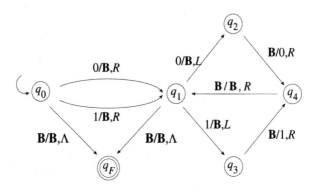

FIGURE 7. The state diagram for our second Turing machine example.

Notice that, as in this example, a state diagram always provides a concise description of a Turing machine. In general, such a description may not be as transparent as a high-level one. If the tape alphabet contains more than a handful of elements, it may be physically impossible to draw the corresponding state diagram clearly.

Referring to any of these descriptions of T, we see that the input 0110 leads to the configuration sequence

$$
\begin{aligned}
(\Lambda, q_0, 0110) &\vdash (\mathbf{B}, q_1, 110) \\
&\vdash (\Lambda, q_3, \mathbf{BB}10) \\
&\vdash (1, q_4, \mathbf{B}10) \\
&\vdash (1\mathbf{B}, q_1, 10) \\
&\vdash (1, q_3, \mathbf{BB}0) \\
&\vdash (11, q_4, \mathbf{B}0) \\
&\vdash (11\mathbf{B}, q_1, 0) \\
&\vdash (11, q_2, \Lambda) \\
&\vdash (110, q_4, \Lambda) \\
&\vdash (110\mathbf{B}, q_1, \Lambda) \\
&\vdash (110\mathbf{B}, q_F, \Lambda).
\end{aligned}
$$

(1.3) Exercises

.1   Let $v \in X^*$, and suppose that there is an infinite sequence

$$
C_0 \equiv (\Lambda, q_0, v), C_1, C_2, \ldots
$$

of admissible configurations such that $C_i \vdash C_{i+1}$ for each i. Must there exist distinct m, n such that $C_m = C_n$ (in which case we say that $\mathcal{M}$ **loops** on the input v)?

.2 Design a Turing machine, with input alphabet $\{0, 1\}$, that shifts a nonempty input word one place to the right, writes a blank in the leftmost cell of the input tape, and parks the read/write head.

.3 Design a Turing machine $\mathcal{M}$ with input alphabet $\{0, 1\}$ and tape alphabet $\{0, 1, \mathbf{B}\}$ such that if $\mathcal{M}$ is started in its start state, with the read/write head on the left and with a string $v \in 0\mathbf{B}^*11^*$ in the leftmost cells of the tape, then $\mathcal{M}$ shifts all the 1's to the left of the tape and parks the read/write head. (This Turing machine will be used in the solution to Exercise (2.7.1).)

.4 Design a Turing machine, with input alphabet $\{0, 1\}$, that executes a **cyclic left shift** by one cell: that is, if the input word is $x_1 \ldots x_N$, with each $x_i \in \{0, 1\}$, then the output word is $x_2 \ldots x_N x_1$ and is written in the leftmost N cells of the input tape. As part of your design, make the Turing machine halt with the read/write head on the left.

.5* Design a Turing machine that duplicates a nonvoid input word over $\{0, 1\}$: that is, if the input word is w, then the output word is ww, with its first symbol in the leftmost cell of the input tape. Make the Turing machine halt with the read/write head on the left.

.6* Design a Turing machine that compares two binary strings, outputs 1 if the strings are equal, outputs 0 if the strings are unequal, and parks the read/write head.

2

Computable Partial Functions

As you may have discovered while doing the exercises at the end of Chapter 1, designing Turing machines to perform particular tasks can be quite an addictive activity. However, that activity is not the object of this book, which is to investigate the *theory,* rather than the practice, of computation. That investigation is based upon the notion of a partial function computed by a Turing machine, to which we now turn our attention.

Let $\mathcal{M} \equiv (\mathcal{Q}, \delta, q_0, q_F)$ be a Turing machine with tape alphabet Y and input alphabet X, and let S be a subset of X^*. We define as follows the **partial function** $\varphi : S \to X^*$ **computed by** $\mathcal{M}$: if $\mathcal{M}$ completes a computation on the input $s \in S$, then $\varphi(s)$ is the output of that computation; otherwise, $\varphi(s)$ is undefined.

For example, consider the computation of the addition function

$$\mathbf{plus} : \mathbf{N}^2 \to \mathbf{N},$$

defined by

$$\mathbf{plus}(m, n) \equiv m + n.$$

We first identify the natural number n with its **unary representation** $\lceil n \rceil$—a string of $n + 1$ terms each equal to 1.[1] We then identify the pair (m, n) of natural numbers with the string $\lceil m \rceil 0 \lceil n \rceil$ in $1\{1\}^*01\{1\}^*$. Thus

$$\mathbf{N} \text{ is identified with } 1\{1\}^*,$$
$$\mathbf{N}^2 \text{ is identified with } 1\{1\}^*01\{1\}^*.$$

The computation of **plus** will be carried out using a **binary Turing machine** $\mathcal{M}$—that is, a Turing machine with input alphabet $\{0, 1\}$ and tape alphabet $\{0, 1, \mathbf{B}\}$; **plus** will be the total partial function from $1\{1\}^*01\{1\}^*$ to $\{0, 1\}^*$ computed by $\mathcal{M}$, and will have values in $1\{1\}^*$.

Here is a high-level description of the behaviour of $\mathcal{M}$ when the initial configuration is $(\Lambda, q_0, \lceil m \rceil 0 \lceil n \rceil)$, with q_0 the start state and $m, n \in \mathbf{N}$. To

---

[1]Normally we shall not distinguish between a natural number n and its unary representation $\lceil n \rceil$. However, there are situations, such as the proof of Theorem (2.8), where we make that distinction in order to avoid confusion.

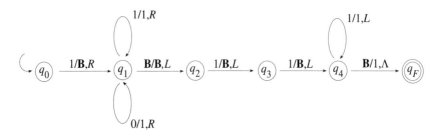

FIGURE 8. A binary Turing machine that computes **plus**.

begin with, $\mathcal{M}$

> writes **B** in the leftmost cell;
> moves right, reading and rewriting 1's, until it reads 0;
> replaces 0 by 1; and
> continues moving right, reading and rewriting 1's, until it reads **B**;

It then

> moves left and deletes 1;
> moves left again and deletes 1;
> continues moving left, reading and rewriting 1's, until it reads **B**;
> writes 1; and
> halts.

$\mathcal{M}$ has then completed a computation, and has $\ulcorner(m+n)\urcorner$—a string of $(m + n + 1)$ terms each equal to 1—in the leftmost cells of the tape, all other tape cells being blank.

Here is the state transition table for $\mathcal{M}$:

	0	1	B
q_0	undefined	$(q_1, \mathbf{B}, R)$	undefined
q_1	$(q_1, 1, R)$	$(q_1, 1, R)$	$(q_2, \mathbf{B}, L)$
q_2	undefined	$(q_3, \mathbf{B}, L)$	undefined
q_3	undefined	$(q_4, \mathbf{B}, L)$	undefined
q_4	undefined	$(q_4, 1, L)$	$(q_F, 1, \Lambda)$
q_F	undefined	undefined	undefined

For the state diagram of $\mathcal{M}$ see Figure 8.

(2.1) Exercises

.1 Design a Turing machine that computes the **empty partial function**
$\epsilon : \{0,1\}^* \to \{0,1\}^*$, where

$$\epsilon(s) \equiv \text{undefined for all } s.$$

.2 Design a Turing machine that computes the partial function **erase** :
$\{0,1\}^* \to \{0,1\}^*$, defined by

$$\textbf{erase}(s) \equiv \Lambda \text{ for all } s.$$

.3 Design Turing machines that compute the **Boolean functions** $\wedge, \vee$,
and $\neg$ on $\{0,1\}^*$, where for each $w \in \{0,1\}^*$,

$$
\begin{aligned}
\wedge(w) &= 1 && \text{if each bit of } w \text{ is } 1, \\
&= 0 && \text{otherwise;} \\
\vee(w) &= 1 && \text{if some bit of } w \text{ is } 1, \\
&= 0 && \text{otherwise;}
\end{aligned}
$$

and $\neg$ is defined inductively by the relations

$$
\begin{aligned}
\neg(\Lambda) &= \Lambda, \\
\neg(w0) &= (\neg w)1, \\
\neg(w1) &= (\neg w)0.
\end{aligned}
$$

.4* Design a Turing machine that computes the multiplication function
times : $\mathbf{N} \times \mathbf{N} \to \mathbf{N}$, defined by

$$\textbf{times}(m, n) \equiv m \times n.$$

.5* Design a Turing machine $\mathcal{M}$ that adds two natural numbers in the
following way: if the numbers have binary representations

$$a \equiv \sum_{k=0}^{N} a_k 2^k \text{ and } b \equiv \sum_{k=0}^{N} b_k 2^k,$$

and if $\mathcal{M}$ is started with the input word $a_0 b_0 a_1 b_1 \dots a_N b_N$ on the left
of the tape, then it completes a computation with output $c_0 c_1 \dots c_K$,
where $a + b = \sum_{j=0}^{K} c_j 2^j$.

The computation of **plus** : $\mathbf{N}^2 \to \mathbf{N}$ preceding Exercises (2.1) typifies our
approach to the computation of partial functions from $\mathbf{N}^n$ to $\mathbf{N}$. Identifying

$$\mathbf{N} \text{ with } 1\{1\}^*,$$
$$\mathbf{N}^n \text{ with } 1\{1\}^* 01\{1\}^* 0 \cdots 01\{1\}^*,$$

we say that a partial function $\varphi : \mathbf{N}^n \to \mathbf{N}$ is **Turing machine computable** if it is the partial function from $\mathbf{N}^n$ to $\mathbf{N}$ computed by some binary Turing machine $\mathcal{M}$. If $m > 1$, we say that a partial function $\varphi : \mathbf{N}^n \to \mathbf{N}^m$ is **Turing machine computable** if the functions $P_k^m \circ \varphi : \mathbf{N}^n \to \mathbf{N}$ ($k = 1, \ldots, m$) are Turing machine computable.

Our definition of *Turing machine computable partial function from $\mathbf{N}^n$ to $\mathbf{N}^m$* is not as restrictive as it may seem: the following lemma will enable us to prove that a partial function $\varphi : \mathbf{N}^n \to \mathbf{N}^m$ is Turing machine computable if it is computed by some Turing machine whose input alphabet includes $\{0, 1\}$.

(2.2) Lemma.   *If $\mathcal{M}$ is a Turing machine whose input alphabet X contains at least two elements, then there is a Turing machine that computes the same partial functions as $\mathcal{M}$ and that has input alphabet X and tape alphabet $X \cup \{\mathbf{B}\}$.*

Proof.   We illustrate the proof by sketching the argument in the case where $X = \{0, 1\}$ and the tape alphabet Y of $\mathcal{M}$ is $\{0, 1, 2, \mathbf{B}\}$. Let

$$\mathcal{Q} \equiv \{q_0, q_1, \ldots, q_F\}$$

be the set of states of $\mathcal{M}$, where q_0 is the start state and q_F the halt state. The idea of the proof is to design a Turing machine $\mathcal{M}'$, with input alphabet X and tape alphabet $X \cup \{\mathbf{B}\}$, that mimics the action of $\mathcal{M}$ on any string over Y by operating on a binary encoding of that string. The set $\mathcal{Q}'$ of states of $\mathcal{M}'$ will include $\mathcal{Q}$ as a proper subset.

When started in its start state, with the input word

$$w \equiv x_1 \ldots x_N \in X^*$$

in the cells on the left of the tape, and with the read/write head against the leftmost cell, $\mathcal{M}'$ first encodes w according to the following scheme:

$$
\begin{aligned}
\text{code}(0) &= 00, \\
\text{code}(1) &= 01, \\
\text{code}(2) &= 10, \\
\text{code}(wx) &= \text{code}(w) \cdot \text{code}(x) \quad (w \in (Y \backslash \{\mathbf{B}\})^*, \ x \in Y \backslash \{\mathbf{B}\}),
\end{aligned}
$$

where the symbol $\cdot$ denotes the concatenation operation on strings. To do so, $\mathcal{M}'$ first moves the entire input string two places to the right and writes blank symbols in the leftmost two cells. It then moves $x_2, \ldots, x_N$ one space right and writes $\mathbf{B}$ in the cell to the right of x_1; moves $x_3, \ldots, x_N$ one space right and writes $\mathbf{B}$ in the cell to the right of x_2; and so on, until the string on the left of the input tape is

$$\mathbf{B}\mathbf{B}x_1\mathbf{B}x_2\mathbf{B}x_3\mathbf{B} \cdots \mathbf{B}x_N$$

and the read/write head is against x_N. Next, the read/write head

> moves left until it has read two successive blanks,
> moves two cells right, and
> reads x_1.

Using states to "memorise" that it read x_1, $\mathcal{M}'$ next

> writes **B**,
> moves left, and
> writes the right bit of code(x_1).

The string on the tape at this stage is

$$\mathbf{B}r\mathbf{BB}x_2\mathbf{B}x_3 \cdots \mathbf{B}x_N$$

where r is the right bit of code(x_1) in the cell being scanned by the read/write head. $\mathcal{M}'$ now

> moves the read/write head three cells to the right,
> reads x_2 (and memorises this by entering the
>   appropriate state),
> writes **B**,
> moves two cells left,
> writes the left bit of code(x_2),
> moves right, and
> writes the right bit of code(x_2).

The string on the tape is now

$$\mathbf{B}r \cdot \text{code}(x_2) \cdot \mathbf{BB}x_3\mathbf{B}x_4 \cdots \mathbf{B}x_N$$

and the read/write head lies against the right bit of code(x_2). Carrying on in this way, we arrive at a configuration with the string

$$\mathbf{B}r \cdot \text{code}(x_2) \cdots \text{code}(x_N)$$

in the leftmost cells of the input tape, and the read/write head against the right bit of code(x_N). $\mathcal{M}'$ then reads the blanks in the three cells to the right of code(x_N); these three blanks indicate that it should complete its task by moving the read/write head left until it reads the leftmost **B**, at which point it writes the left bit of code(x_1) and enters the state q_0.[2]

---

[2]Note that at each stage the module stores relevant information by means of the state the machine enters. For example, when the read/write head reads x_1 prior to moving left and writing the right bit of code(x_1), the machine enters a different state for each of the possible values of x_1.

Renaming the states of $\mathcal{M}'$ used so far, we can ensure that none of those states except q_0 belongs to $\mathcal{Q}$.

We next arrange for $\mathcal{M}'$ to imitate the passage from one configuration of $\mathcal{M}$ to the next. If

$$\delta : \mathcal{Q} \times Y \to \mathcal{Q} \times Y \times \{L, R, \Lambda\}$$

is the state transition function of $\mathcal{M}$, $\mathcal{M}'$ imitates the transition represented by $\delta(q, y) = (q', y', D)$ as follows. Suppose that $\mathcal{M}'$ is in the state q, that $y \in \{0, 1, 2\}$, and that the read/write head of $\mathcal{M}'$ lies against the left bit of code(y). $\mathcal{M}'$ reads that bit, moves right (using states as memory), and registers that the two bits just read form code(y); it then moves left, writes

B if $y' = \mathbf{B}$,
the left bit of code(y') if $y' \in Y \backslash \{\mathbf{B}\}$,

and uses states to remember q', y', and D. $\mathcal{M}'$ then moves right, reads the right bit of code(y), writes

B if $y' = \mathbf{B}$,
the right bit of code(y') if $y' \in Y \backslash \{\mathbf{B}\}$,

moves

three places left if $D = L$,
one place left if $D = \Lambda$,
one place right if $D = R$,

and passes to the state q'.

If, on the other hand, $y = \mathbf{B}$ and $\mathcal{M}'$ is in the state q with the read/write head against $\mathbf{B}$, then $\mathcal{M}'$ proceeds as follows. It reads $\mathbf{B}$, moves right (using states as memory), and registers that the symbol against the read/write head is $\mathbf{B}$. $\mathcal{M}'$ then moves left, writes

B if $y' = \mathbf{B}$,
the left bit of code(y') if $y' \in Y \backslash \{\mathbf{B}\}$,

and uses states to remember q', y', and D. $\mathcal{M}$ then

moves right,
reads **B**,
writes
    B if $y' = \mathbf{B}$,
    the right bit of code(y') if $y' \in Y \backslash \{\mathbf{B}\}$,
moves
    three places left if $D = L$,
    one place left if $D = \Lambda$,
    one place right if $D = R$,

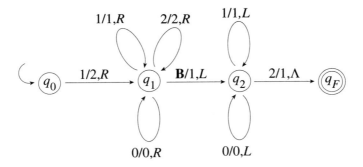

FIGURE 9. The Turing machine $\mathcal{M}$ in Exercise (2.4.1).

and passes to the state q'. Finally, if $q' = q_N$, the halt state of $\mathcal{M}$, we require $\mathcal{M}'$ to halt on entering q'.   □

(2.3) Proposition.   *If φ is a partial function from $\mathbf{N}^n$ to $\mathbf{N}$ that is computed by some Turing machine whose input alphabet contains $\{0, 1\}$, then φ is Turing machine computable.*

Proof.   To make sense of this proposition we must remember that we are identifying each natural number with its unary representation. Given a Turing machine $\mathcal{M}$, with input alphabet $X \supset \{0, 1\}$ and start state q_0, that computes φ, we construct a binary Turing machine $\mathcal{M}'$ as follows. First, we delete from the state diagram of $\mathcal{M}$ any arrows representing transitions of the form

$$\delta(q_0, y) = (q', y', D)$$

with $y \in X \backslash \{0, 1\}$. Next, we restrict the input alphabet to $\{0, 1\}$, and use Lemma (2.2) to construct a binary Turing machine $\mathcal{M}'$ that computes the same partial functions from $\{0, 1\}^*$ to $\{0, 1\}^*$ as does $\mathcal{M}$. In particular, when started in its start state with with the input word $k_0 0 k_1 0 \ldots 0 k_n$ on the left of the tape, where $k_i \in \mathbf{N}$ for each i, $\mathcal{M}'$ completes a computation with output $\varphi(k_1, \ldots, k_n)$.   □

(2.4) Exercises

.1*   Carry out the construction in the proof of Lemma (2.2) to design a binary Turing machine $\mathcal{T}$ that computes the same partial functions as the Turing machine $\mathcal{M}$ with input alphabet $\{0, 1\}$, tape alphabet $\{0, 1, 2, \mathbf{B}\}$, and the state diagram described in Figure 9. (This Turing machine adds 1 to the rightmost projection of an input string from $\mathbf{N}^n$.)

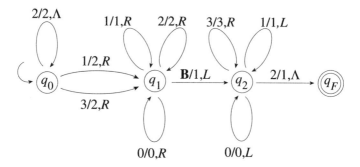

FIGURE 10. The Turing machine $\mathcal{M}$ in Exercise (2.4.2).

.2   Let φ be the partial function from $\mathbf{N}$ to $\mathbf{N}$ computed by the Turing machine $\mathcal{M}$ with input alphabet $\{0, 1, 2\}$, tape alphabet $\{0, 1, 2, 3, \mathbf{B}\}$, and the state diagram described in Figure 10. Carry out the construction in the proof of Proposition (2.3) to design a binary Turing machine that computes φ.

.3*   Fill in the details of the proof of Proposition (2.3).

The year 1936 marks the beginning of the modern era of the theory of computation, with the introduction of three mathematically precise notions attempting to capture the informal idea of a computable partial function from $\mathbf{N}^n$ to $\mathbf{N}$. These notions are Turing machines, Kleene's partial recursive functions [20, 23], and Church's lambda calculus [2]. It was shown subsequently that these three and all other attempts to characterise computable partial functions give rise to the same class of computable partial functions from $\mathbf{N}^n$ to $\mathbf{N}$—namely, those that are Turing machine computable; see [23], Chapter 1.

Nevertheless, partial recursive functions and the lambda calculus are of interest in their own right. Both are significant in the theory and practice of programming languages. In particular, the lambda calculus (which we shall not discuss further) underpins the language LISP. On the other hand, the concepts and methods of recursive function theory have permeated mathematics and logic to such an extent that all mathematicians should be aware of what a recursive function looks like; for this reason we now make a short detour to look more closely at those functions.

We begin with the class of **base functions**, which comprises

- the natural numbers $0, 1, 2, \ldots$ (considered as functions of zero variables);

- the **zero function** $\mathbf{0} : \mathbf{N} \to \mathbf{N}$, defined by $\mathbf{0}(n) \equiv 0$;

- the **successor function scsr** : $\mathbf{N} \to \mathbf{N}$, defined by $\mathbf{scsr}(n) \equiv n+1$;

- the **projection functions** $P_j^n : \mathbf{N}^n \to \mathbf{N}$, defined by

$$P_j^n(k_1, \ldots, k_n) \equiv k_j \text{ for all } k_1, \ldots, k_n \text{ in } \mathbf{N}.$$

For $n \geq 1$, the partial function $\varphi : \mathbf{N}^n \to \mathbf{N}$ is obtained from the partial functions $\psi : \mathbf{N}^{n-1} \to \mathbf{N}$ and $\theta : \mathbf{N}^{n+1} \to \mathbf{N}$ by **primitive recursion** if for all $k, k_2, \ldots, k_n$ in $\mathbf{N}$,

$$\varphi(0, k_2, \ldots, k_n) = \psi(k_2, \ldots, k_n)$$

and

$$\varphi(k+1, k_2, \ldots, k_n) = \theta(k, \varphi(k, k_2, \ldots, k_n), k_2, \ldots, k_n).$$

(Recall that in such a definition it is understood that the left-hand side is defined if and only if the right-hand side is defined.) In particular, the total function $f : \mathbf{N} \to \mathbf{N}$ is obtained from the constant $c \in \mathbf{N}$ and the total function $h : \mathbf{N}^2 \to \mathbf{N}$ by primitive recursion if $f(0) = c$ and $f(k+1) = h(k, f(k))$ $(k \in \mathbf{N})$.

The set $\mathcal{P}$ of **primitive recursive functions** over $\mathbf{N}$ is defined inductively by the following conditions, where $m, n \geq 1$:

- $\mathcal{P}$ contains all the base functions.

- If $g : \mathbf{N}^m \to \mathbf{N}$ and $h_k : \mathbf{N}^n \to \mathbf{N}$ $(k = 1, \ldots, m)$ belong to $\mathcal{P}$, then the composite function $g \circ (h_1, \ldots, h_m) : \mathbf{N}^n \to \mathbf{N}$ belongs to $\mathcal{P}$.

- If the functions $g : \mathbf{N}^{n-1} \to \mathbf{N}$ and $h : \mathbf{N}^{n+1} \to \mathbf{N}$ belong to $\mathcal{P}$, then so does the function $f : \mathbf{N}^n \to \mathbf{N}$ obtained from g and h by primitive recursion.

For example, the functions **plus** and **times** on $\mathbf{N}^2$ (introduced earlier in this chapter) are primitive recursive: for

$$
\begin{aligned}
\mathbf{plus}(0, k) &= P_1^1(k), \\
\mathbf{plus}(j+1, k) &= \mathbf{scsr} \circ P_2^3(j, \mathbf{plus}(j, k), k), \\
\mathbf{times}(0, k) &= 0, \\
\mathbf{times}(j+1, k) &= \mathbf{plus} \circ (P_2^3, P_3^3)(j, \mathbf{times}(j, k), k).
\end{aligned}
$$

(2.5) Exercise

Prove that $\mathcal{P}$ is the set of all functions obtained from the base functions by finitely many applications of composition and primitive recursion.

The application of recursion or composition to total functions on $\mathbf{N}^n$ always produces total functions. In order to construct from the elements of $\mathcal{P}$ an appropriate class of *partial* functions on $\mathbf{N}^n$, we introduce one more method of obtaining new functions from old.

An element $(k_1, \ldots, k_n)$ of $\mathbf{N}^n$ is **admissible for minimisation** relative to the partial function $\psi : \mathbf{N}^{n+1} \to \mathbf{N}$ if

$$
\mathcal{D}(k_1, \ldots, k_n) \equiv \{m \in \mathbf{N} : (i, k_1, \ldots, k_n) \in \mathrm{domain}(\psi) \ (0 \leq \\
i \leq m) \text{ and } \psi(m, k_1, \ldots, k_n) = 0\}
$$

is nonempty. The partial function $\varphi : \mathbf{N}^n \to \mathbf{N}$ is **obtained from ψ by minimisation** if the domain of φ is the set of those $(k_1, \ldots, k_n) \in \mathbf{N}^n$ that are admissible for minimisation relative to ψ, and

$$
\varphi(k_1, \ldots, k_n) = \min \mathcal{D}(k_1, \ldots, k_n)
$$

for each $(k_1, \ldots, k_n) \in \mathrm{domain}(\varphi)$. In that case we write

$$
\varphi(k_1, \ldots, k_n) \equiv \min k \, [\psi(k, k_1, \ldots, k_n) = 0].
$$

The set $\mathcal{R}$ of **partial recursive functions** over $\mathbf{N}$ is defined inductively by the following four conditions, where $m, n \geq 1$:

- $\mathcal{R}$ contains all the base functions.

- If $\varphi : \mathbf{N}^m \to \mathbf{N}$ and $\psi_k : \mathbf{N}^n \to \mathbf{N}$ $(k = 1, \ldots, m)$ belong to $\mathcal{R}$, then the composite function $\varphi \circ (\psi_1, \ldots, \psi_m) : \mathbf{N}^n \to \mathbf{N}$ belongs to $\mathcal{R}$.

- If $\psi : \mathbf{N}^{n-1} \to \mathbf{N}$ and $\theta : \mathbf{N}^{n+1} \to \mathbf{N}$ belong to $\mathcal{R}$, then the partial function $\varphi : \mathbf{N}^n \to \mathbf{N}$ obtained from ψ and θ by primitive recursion belongs to $\mathcal{R}$.

- If the partial function $\psi : \mathbf{N}^{n+1} \to \mathbf{N}$ belongs to $\mathcal{R}$, then so does the partial function $\varphi : \mathbf{N}^n \to \mathbf{N}$ obtained from ψ by minimisation.

Thus (cf. Exercise (2.5)) $\mathcal{R}$ is the set of all partial functions obtained from the base functions by finitely many applications of composition, primitive recursion, and minimisation. Clearly, $\mathcal{P} \subset \mathcal{R}$.

(2.6) Exercises

.1   Prove that the **factorial function** $n \mapsto n!$ is primitive recursive on $\mathbf{N}$.

.2   Prove that the function **power** : $\mathbf{N}^2 \to \mathbf{N}$, defined by

$$
\mathbf{power}(m, n) \equiv n^m,
$$

is primitive recursive. Does

$$\mathbf{power}'(m, n) \equiv m^n$$

define a primitive recursive function? What about the function $n \mapsto n^m$, where $m \in \mathbf{N}$ is fixed?

.3   Prove that the **cutoff subtraction** function, defined by

$$\begin{aligned}\mathbf{cutoff}(m, n) \;\;&=\;\; m - n \quad \text{if } m \geq n, \\ &=\;\; 0 \qquad\quad \text{otherwise,}\end{aligned}$$

is primitive recursive on $\mathbf{N}^2$. (*Hint*: First prove that the function $m \mapsto$ **cutoff**$(m, 1)$ is primitive recursive on $\mathbf{N}$.) Prove also that $(m, n) \mapsto |m - n|$ is a primitive recursive function, where $|\cdot|$ denotes absolute value.

.4   Prove that the partial function $\mathbf{sqrt} : \mathbf{N} \to \mathbf{N}$, defined by

$$\begin{aligned}\mathbf{sqrt}(n) \;\;&=\;\; \sqrt{n} \qquad\quad\; \text{if } n \text{ is a perfect square,} \\ &=\;\; \text{undefined} \quad \text{otherwise,}\end{aligned}$$

belongs to $\mathcal{R}$. (Recall that $\sqrt{\cdot}$ denotes the nonnegative square root.)

.5*   Let $\varphi : \mathbf{N}^2 \to \mathbf{N}$ and $\psi : \mathbf{N} \to \mathbf{N}$ be partial recursive functions, and for each n define

$$S(n) \equiv \{k \in \mathbf{N} : \varphi(n, k) \leq \psi(n)\}.$$

Prove that

$$\begin{aligned}\theta(n) \;\;&=\;\; \min S(n) \quad\; \text{if } S(n) \text{ is nonempty,} \\ &=\;\; \text{undefined} \quad \text{otherwise}\end{aligned}$$

defines a partial recursive function from $\mathbf{N}$ to $\mathbf{N}$. We often write

$$\theta(n) = \min k[\varphi(n, k) \leq \psi(n)].$$

The following exercises take much of the sting out of the proof that every function in $\mathcal{R}$ is Turing machine computable. Note that we take each natural number, considered as a function of zero variables, to be computable *by convention*.

(2.7) Exercises

.1   Prove that each of the base functions is Turing machine computable.

.2   Let $\psi : \mathbf{N}^m \to \mathbf{N}$ and $\theta_1, \ldots, \theta_m : \mathbf{N}^n \to \mathbf{N}$ be Turing machine computable partial functions. Prove that the composite function $\varphi \equiv \psi \circ (\theta_1, \ldots, \theta_m)$ is a Turing machine computable partial function from $\mathbf{N}^n$ to $\mathbf{N}$.

.3   Let ψ be a Turing machine computable partial function from $\mathbf{N}^{n+1}$ to $\mathbf{N}$. Prove that the partial function obtained from ψ by minimisation is Turing machine computable.

(2.8) Theorem.   *Every partial recursive function $\varphi : \mathbf{N}^n \to \mathbf{N}$ is Turing machine computable.*

Proof.   Given Turing machine computable partial functions

$$\psi : \mathbf{N}^{n-1} \to \mathbf{N}, \quad \theta : \mathbf{N}^{n+1} \to \mathbf{N},$$

define the partial function $\varphi : \mathbf{N}^n \to \mathbf{N}$ recursively by

$$
\begin{aligned}
\varphi(i, u) &= \psi(u) && \text{if } i = 0, \\
&= \theta(i - 1, \varphi(i - 1, u), u) && \text{if } i \geq 1.
\end{aligned}
$$

In view of Exercises (2.7), we need only describe a binary Turing machine $\mathcal{M}$ that computes φ; for simplicity, we take the case $n = 2$. Let i, j be natural numbers, and consider the behaviour of $\mathcal{M}$ when it is started in the start state with the input string $\lceil i \rceil 0 \lceil j \rceil$ on the left of the tape. $\mathcal{M}$ begins by reading and rewriting 1 in the leftmost cell, moving one square right, and entering a special "checking" state q. If it then reads 0 in the second cell on the left (in which case $i = 0$), it calls a module that writes $\lceil \psi(j) \rceil$ in the leftmost cells, leaves all the other cells blank, and parks the head. If, on the other hand, $\mathcal{M}$, in the state q, reads 1 in the second cell on the left (in which case $i \geq 1$), it moves the input string $\lceil i \rceil 0 \lceil j \rceil$ one place to the right, writes $\mathbf{B}$ in the leftmost cell, and writes $00 \lceil (i - 1) \rceil 0 \lceil j \rceil$ on the right of $\lceil j \rceil$. Leaving each of the zeroes unchanged in position on the tape, $\mathcal{M}$ then calls a module that

> leaves the tape unchanged at and to the left of the rightmost copy of $\lceil j \rceil$,
> writes $0010 \lceil \psi(j) \rceil$ on the right of that copy of $\lceil j \rceil$,
> leaves the read/write head against the cell immediately to the right of the leftmost instance of 00, and
> enters a special state q_1.

Now assume that $\mathcal{M}$ is in the state q_1 with the read/write head against the cell immediately to the right of the leftmost instance of 00; that the leftmost cells of the tape contain

$$\mathbf{B}^{\lceil i \rceil} 0^{\lceil j \rceil} 00 w s 0^{\lceil j \rceil} 00^{\lceil k \rceil} 0^{\lceil \varphi(k,j) \rceil}$$

where $0 \le k < i$, $w \in \{1\}^*$, $s \in \{\mathbf{B}\}^*$, and the number of cells occupied by ws is i; and that the rest of the tape is blank. $\mathcal{M}$ then calls a Turing machine module $\mathcal{T}$ that first examines the left symbol of ws. If that bit is 1—so that $|s| < i$—then $\mathcal{T}$ changes the rightmost bit of w to $\mathbf{B}$; thus $ws = w's'$, where

$$w' \in \{1\}^*, \ s' \in \{\mathbf{B}\}^*,$$

$|w'| = |w| - 1$, $|s'| = |s| + 1$, and $|w's'| = |ws| = i$.

$\mathcal{T}$ then writes $0^{\lceil j \rceil}$ on the right of the string $\lceil \varphi(k,j) \rceil$ on the right of the tape, and places the read/write head against the cell to the right of the rightmost instance of 00. Next, $\mathcal{T}$ calls a module that, leaving the tape unchanged to the left of the string $\lceil k \rceil 0 \lceil \varphi(k,j) \rceil 0 \lceil j \rceil$,

> replaces that string with $\lceil (k+1) \rceil 0 \lceil \theta(k, \varphi(k,j), j) \rceil$,
> leaves the read/write head against the cell immediately to the
>   right of the leftmost instance of 00, and
> enters the special state q_1.

This completes the action of the module $\mathcal{T}$ and leaves $\mathcal{M}$ ready for a further call of that same module.

On the other hand, if, on its initial examination of the left symbol of ws, $\mathcal{T}$ discovers that that symbol is $\mathbf{B}$, then $w = \Lambda$ and $|s| = i$; in that case, $\mathcal{T}$ (and therefore $\mathcal{M}$) moves right, replacing each $\mathbf{B}$ that it reads by 0, until it reaches the rightmost instance of 0. It then

> copies the string $\varphi(i,j)$ from the right to the far left of the tape,
>   leaving all other tape cells blank, and
> halts with the read/write head parked on the left (cf. Exercise (1.3.3)).

($\mathcal{M}$ recognises the leftmost cell on the tape by the blank it deposited there early in its execution.) In that case, $\mathcal{M}$ has completed a computation with output $\varphi(i,j)$.   □

You should note that as long as Exercises (2.7) have been carried out correctly, Theorem (2.8) provides an *effective* method of obtaining a binary Turing machine that computes a given partial recursive function from $\mathbf{N}^n$ to $\mathbf{N}$.

Designing binary Turing machines to carry out even simple computational tasks such as the addition of two integers can be an intricate business; indeed, in its need for careful attention to fine details, Turing machine design is reminiscent of machine language programming. Fortunately, in theoretical studies of computability it is customary to believe the over-whelming mass of evidence that supports the **Church-Markov-Turing thesis**[3]:

> *A partial function $\varphi : \mathbf{N}^n \to \mathbf{N}$ is computable (in any accepted informal sense) if and only if it is computable by some binary Turing machine—that is, if and only if $\varphi = \varphi_k$ for some k.*

By accepting this thesis, as we shall do from now on, we are able to dispense with a formal proof of Turing machine computability, provided that the partial function under consideration is clearly computable in some informal sense.

For example, if $\varphi : \mathbf{N} \to \mathbf{N}$ is a Turing machine computable partial function, then in order to prove that the composite partial function $\varphi \circ \varphi$ is Turing machine computable, we first note the following informal algorithm for computing $\varphi \circ \varphi(k)$: choose a binary Turing machine $\mathcal{M}$ that computes φ, and run $\mathcal{M}$ on the input $\lceil k \rceil$; if $\mathcal{M}$ completes a computation, run $\mathcal{M}$ again, this time with the input $\lceil \varphi(k) \rceil$. We then invoke the Church-Markov-Turing thesis.

We shall use such informal arguments, with an (often unstated) appeal to the Church-Markov-Turing thesis, throughout the remainder of this book. Nevertheless, there will be situations where a given partial function is not obviously computable in any informal sense; in such circumstances we shall confirm the function's computability by describing a Turing machine that computes it.

It is important to realise that the Church-Markov-Turing thesis is not susceptible of proof: it is an *unsubstantiable claim* that all notions, formal and informal, of a computable partial function from $\mathbf{N}^n$ to $\mathbf{N}$ are equivalent to the formal notion of a Turing machine computable partial function. Our willingness to accept the Church-Markov-Turing thesis is based on

- the fact, mentioned above, that all attempts to formalise the intuitive notion of a computable partial function from $\mathbf{N}^n$ to $\mathbf{N}$ have led to the same class of functions, and

- the absence of any convincing example of a computable partial function that is not Turing machine computable.[4]

Theorem (2.8) proves part of the identification of the partial recursive functions with the Turing machine computable functions; for the remainder of that proof, and proofs of other such identifications, see Chapter 1 of [23].

---

[3]This is commonly known as *Church's thesis;* but our name for it reflects more accurately its origins.

[4][19] contains an argument against the Church-Markov-Turing thesis; this is discussed briefly on page 142 of [25].

We shall not pursue those proofs here, since we are primarily interested in the consequences of the Church-Markov-Turing thesis, rather than the detailed justification of the thesis itself.

We end this chapter with an observation about the classes $\mathcal{P}$ and $\mathcal{R}$. Although the extension from $\mathcal{P}$ to $\mathcal{R}$ was made in order to accommodate partial functions within Kleene's theory of computation, *there are total partial recursive functions over* $\mathbf{N}$ *that are not primitive recursive*. A standard example of such a function is **Ackermann's function** $A : \mathbf{N}^2 \rightarrow \mathbf{N}$, defined by the equations

$$
\begin{aligned}
A(0, n) &\equiv n + 1, \\
A(m + 1, 0) &\equiv A(m, 1), \\
A(m + 1, n + 1) &\equiv A(m, A(m + 1, n)).
\end{aligned}
$$

It is intuitively clear that A is computable, and hence, by the Church-Markov-Turing thesis, that it is partial recursive. Thus the function $A' : \mathbf{N} \rightarrow \mathbf{N}$ given by

$$
A'(n) \equiv A(n, n)
$$

is a total partial recursive function on $\mathbf{N}$. But, as is shown by an involved argument that can be found on pages 11-21 of [9], to each primitive recursive function $f : \mathbf{N} \rightarrow \mathbf{N}$ there correspond m and k such that $A(m, n) > f(n)$ whenever $n \geq k$. It follows from this that A', and therefore A, cannot be primitive recursive; see Exercise (2.9.2).

(2.9) Exercises

.1   Prove that Ackermann's function is a total function on $\mathbf{N}^2$.

.2*   Prove each of the following statements about Ackermann's function.

   (i) $A(m, n) > n$.

   (ii) $A(m, n + 1) > A(m, n)$.

   (iii) $A(m + 1, n) > A(m, n)$.

   (iv)   If $n \geq 2$, then $A(m, A(m, n)) > 2A(m, n)$.

   (v) $A(m + 1, n) = A(m, A(m, \ldots, A(m, 1) \ldots))$, where A appears $n + 1$ times on the right-hand side.

   Use (iii) and the observation in the paragraph preceding these exercises to show that to each primitive recursive function $f : \mathbf{N} \rightarrow \mathbf{N}$ there corresponds a natural number n such that $A'(n) > f(n)$.

.3   Prove that

$$
A(4, n) = 2^{2^{\cdot^{\cdot^{\cdot^2}}}} - 3,
$$

where there are $n + 3$ instances of the symbol 2 on the right-hand side. (*Hint:* First find expressions for $A(1, n)$, $A(2, n)$, and $A(3, n)$.)

3

Effective Enumerations

Is every subset of $\mathbf{N}$ the domain of some computable partial function? If not, can we characterise those subsets of $\mathbf{N}$ that are domains of computable partial functions?

We begin this chapter by introducing the fundamental notions of *effective enumeration* and *recursively enumerable set* and applying them to answer the latter question affirmatively (the discussion of the former being deferred until Chapter 4). We then define recursive sets and describe an algorithm that enables us to identify Turing machines of a certain type as the elements of a recursive set of natural numbers. Taken with the Church-Markov-Turing thesis, this provides us with an effective enumeration of the set of all computable partial functions from $\mathbf{N}^n$ to $\mathbf{N}$. Using the language of sonata form, we might say that this enumeration is the transition from the exposition in Chapters 1 and 2 to the development of the subject of computability; that development begins with the *s-m-n* theorem towards the end of this chapter.

Let X be a set, and S a subset of X. We say that S is **countable** if either $S = \emptyset$ or there exists a total function f from $\mathbf{N}$ onto S; in the latter case the function f is called an **enumeration** of S. Such an enumeration it is often described by, and identified with, the list

$$f(0), f(1), \ldots$$

of its values.

Now suppose that we have defined the notion of a computable partial function from $\mathbf{N}$ into X. By an **effective enumeration** of S we mean a total *computable* function f from $\mathbf{N}$ onto S, which is then said to be **effectively enumerable** (by f). Of particular importance is the case $X = \mathbf{N}$, when we also say that f is a **recursive enumeration** of S and that S is **recursively enumerable** (by f).

Following convention, we also call the empty subset of X *effectively enumerable* or, in the case $X = \mathbf{N}$, *recursively enumerable*.

(3.1) Exercise

Prove that the union and the intersection of two recursively enumerable subsets of $\mathbf{N}$ are recursively enumerable.

(3.2) Proposition.   *If S is a recursively enumerable subset of $\mathbf{N}$, then the partial function $\varphi : \mathbf{N} \to \mathbf{N}$ defined by*

$$\begin{aligned} \varphi(n) \quad &= \quad 1 \qquad\qquad\quad\; \textit{if } n \in S, \\ &= \quad \text{undefined} \quad\; \textit{if } n \notin S \end{aligned}$$

is computable.

Proof.   If S is empty, then φ is the empty partial function on $\mathbf{N}$, which is computable by Exercise (2.1.1); so we may suppose that S is nonempty. Then there exists a total computable function f from $\mathbf{N}$ onto S. The basic idea underlying the construction of a Turing machine $\mathcal{M}$ that computes φ is simple: we compare the input n with $f(0), f(1), f(2), \ldots$ in turn, and output 1 if we come across $k \in \mathbf{N}$ such that $n = f(k)$. More precisely, let $\mathcal{M}$ have input alphabet $\{0, 1\}$ and tape alphabet $\{0, 1, \mathbf{B}\}$. With initial configuration (Λ, q_0, n), where q_0 is its start state and n is (the unary form of) a nonnegative integer, $\mathcal{M}$ first writes $\mathbf{B}$ as a left-end marker in the leftmost cell. Without affecting the remaining units of n, $\mathcal{M}$ then writes the string $0\mathbf{B}0n$ in the cells on the right of the tape, and enters a special state q, with the read/write head scanning the rightmost instance of 0.

Now let k be either $\mathbf{B}$ or (the unary form of) a natural number. Suppose that the leftmost cells of the tape contain the string $\mathbf{B}n'0k0n$ and that all other cells are blank, where n' is the string formed by deleting the leftmost unit of n. Suppose also that $\mathcal{M}$ is in the state q, with the read/write head scanning the rightmost instance of 0. $\mathcal{M}$ then calls a Turing machine module $\mathcal{M}'$ that, without changing the content of the cells at and to the left of n,

> copies $0k$ to the right of n,
> calculates $f(k)$ and writes $0f(k)$ in the cells to the right of n, and
> checks whether the unary strings n and $f(k)$ are equal.

If $n = f(k)$, $\mathcal{M}$ then writes blank symbols in every tape cell except the leftmost one, where it writes 1; finally, it passes to its halt state and parks the read/write head. If $n \neq f(k)$, then, without changing the content of the cells at and to the left of the leftmost 0, $\mathcal{M}$

> deletes the string on the right of k,
> calculates $k + 1$,
> writes (the unary form of) $k + 1$, followed by $0n$,
>   on the right of the leftmost 0,
> moves left until it reads the rightmost 0,
> rewrites 0, and
> passes to the state q.

The construction of $\mathcal{M}$ completes the proof that φ is a computable partial function on $\mathbf{N}$.   □

Let $C \equiv (u, q, v)$ be a configuration of a Turing machine $\mathcal{M}$, q_0 the start state of $\mathcal{M}$, and w a string over the input alphabet of $\mathcal{M}$. We say that $\mathcal{M}$ **reaches the configuration** C **in** k **steps on the input** w if $(\Lambda, q_0, w) \vdash^k C$. If also q is the halt state of $\mathcal{M}$, we say that $\mathcal{M}$ **halts in** k **steps on the input** w. If, for some $k \leq n$, $\mathcal{M}$ halts in k steps on the input w, we say that $\mathcal{M}$ **halts in at most** n **steps on the input** w.

These notions prepare us for the proof of a very important characterization of recursively enumerable sets.

(3.3) Theorem.   *A subset S of $\mathbf{N}$ is recursively enumerable if and only if it is the domain of a computable partial function on $\mathbf{N}$.*

Proof.   In view of Proposition (3.2), we need only consider the sufficiency of the stated condition. Accordingly, consider a subset S of $\mathbf{N}$ that is the domain of a computable partial function $\varphi : \mathbf{N} \to \mathbf{N}$. We may assume that S is nonempty. Fixing an element a of S, let $\mathcal{M}$ be a Turing machine that computes φ, and define a total computable function h from $\mathbf{N}^2$ onto S as follows:

$$
\begin{aligned}
h(i, j) \quad &= \quad i \quad \text{if } \mathcal{M} \text{ completes a computation in at} \\
&\qquad\qquad \text{most } j \text{ steps on the input } i, \\
&= \quad a \quad \text{otherwise.}
\end{aligned}
$$

If we now follow the arrows through the diagram

$$
\begin{array}{llll}
h(0,0) \longrightarrow & h(0,1) & h(0,2) \longrightarrow & h(0,3) \ \cdots \\
\quad\quad\quad \swarrow & \quad\quad \nearrow & \quad\quad \swarrow & \\
h(1,0) & h(1,1) & h(1,2) \ \cdots & \\
\quad \downarrow \quad\quad \nearrow & \quad\quad \swarrow & & \\
h(2,0) & h(2,1) \quad\quad \cdots & & \\
\quad\quad\quad \swarrow & & & \\
h(3,0) \quad\quad \cdots & & & \\
\quad \downarrow & & & \\
\quad \vdots & & &
\end{array}
$$

we obtain an effective enumeration

$$
h(0,0), h(0,1), h(1,0), h(2,0), h(1,1), h(0,2), \ldots
$$

of S; whence S is recursively enumerable. $\square$

(3.4) Exercises

.1   Why is the function h defined in the above proof computable?

.2   Prove that a subset of $\mathbf{N}$ is recursively enumerable if and only if it is the range of a computable partial function from $\mathbf{N}$ to $\mathbf{N}$.

We call a subset S of $\mathbf{N}$ **recursive** if its **characteristic function** $\chi_S : S \to \mathbf{N}$, defined by

$$\chi_S(n) \;\; = \;\; 1 \quad \text{if } n \in S,$$
$$= \;\; 0 \quad \text{if } n \notin S,$$

is a total computable function on $\mathbf{N}$. Thus S is recursive if and only if there is an algorithm for deciding whether any given element of $\mathbf{N}$ belongs to S. For example, $\mathbf{N}$ is recursive, since each element of $\mathbf{N}$ belongs to $\mathbf{N}$; and the empty subset $\emptyset$ of $\mathbf{N}$ is recursive, since each element of $\mathbf{N}$ is not in $\emptyset$.

(3.5) Exercises

.1   Prove that if S is an infinite recursive subset of $\mathbf{N}$, then there exists a strictly increasing (and therefore one-one) total computable function f from $\mathbf{N}$ onto S. Prove also that

$$\varphi(n) \;\; = \;\; f^{-1}(n) \quad \text{if } n \in S,$$
$$= \;\; \text{undefined} \quad \text{if } n \notin S$$

defines a computable partial function $\varphi : \mathbf{N} \to \mathbf{N}$.

.2   Prove that a recursive subset of $\mathbf{N}$ is recursively enumerable.

.3   Prove that a subset S of $\mathbf{N}$ is recursively enumerable if and only if there exists a computable partial function $\varphi : \mathbf{N} \to S$ whose domain is a recursive subset of $\mathbf{N}$ and whose range is S.

Proposition (3.2) and Theorem (3.3) enable us to shed more light on the distinction between recursive and recursively enumerable sets: for they show that a subset S of $\mathbf{N}$ is recursively enumerable if and only if the partial function $\varphi : \mathbf{N} \to \mathbf{N}$ defined by

$$\varphi(n) \;\; = \;\; 1 \quad \text{if } n \in S,$$
$$= \;\; \text{undefined} \quad \text{if } n \notin S$$

is computable. On the other hand, S is recursive if and only if we can replace *undefined* by 0 in the definition of φ and still obtain a computable (now total) function on $\mathbf{N}$. There remains, however, the possibility that this distinction is illusory and that every recursively enumerable set is recursive. We shall see in the next chapter that this is not the case; meanwhile, we turn our attention to the encoding of Turing machines as natural numbers.

A binary Turing machine is said to be **normalised** if there exists a natural number N such that the Turing machine has states $0, 1, 2, \ldots, N$, with initial state 0 and halt state N. Note that, in contrast to our usual practice of identifying natural numbers with their unary representations, we describe a state of a normalised binary Turing machine by its minimal *decimal* representation—that is, the decimal representation with the fewest digits. We denote by $\mathcal{N}$ the set of all normalised binary Turing machines.

We now describe an algorithmic encoding procedure which will enable us to identify $\mathcal{N}$ with a recursive subset of $\mathbf{N}$. To begin with, we set up an encoding of the decimal digits and the symbols $\mathbf{B}$, $\perp$ ("undefined"), L (left move), R (right move), Λ (no move), and $/$ (auxiliary separator), as follows:

σ	code(σ)
0	10000
1	10001
2	10010
3	10011
4	10100
5	10101
6	10110
7	10111
8	11000
9	11001
$\mathbf{B}$	11010
$\perp$	11011
L	11100
R	11101
Λ	11110
$/$	11111

If $m_k \ldots m_1 m_0$ is the minimal decimal form of a natural number m, we define
$$\text{code}(m) \equiv \text{code}(m_k) \cdots \text{code}(m_1) \cdot \text{code}(m_0).$$
For each triple $t \equiv (q, y, D)$ with $q \in \mathbf{N}$, $y \in \{0, 1, \mathbf{B}\}$, and $D \in \{L, R, \Lambda\}$, we define
$$\text{code}(t) \equiv \text{code}(q) \cdot \text{code}(/) \cdot \text{code}(y) \cdot \text{code}(/) \cdot \text{code}(D).$$

Now consider any normalised binary Turing machine
$$\mathcal{M} \equiv (\mathcal{Q}, \delta, q_0, q_F),$$
where, for some natural number N, $\mathcal{Q} = \{0, 1, 2, ..., N\}$, $q_0 = 0$, and $q_F = N$. $\mathcal{M}$ is completely specified by the integer N and the values $\delta(i, j)$ ($0 \leq i \leq N - 1$) of the transition function
$$\delta : \{0, 1, ..., N\} \times \{0, 1, \mathbf{B}\} \to \{0, 1, \ldots, N\} \times \{0, 1, \mathbf{B}\} \times \{L, R, \Lambda\}. \quad (3.1)$$

To encode $\mathcal{M}$, first form the string

$$N/\delta(0,0)/\delta(0,1)/\delta(0,\mathbf{B})/\delta(1,0)/\delta(1,1)/\delta(1,\mathbf{B})/ \ \dots \ /\delta(N-1,\mathbf{B}) \quad (3.2)$$

and then encode it by concatenating the codings of its various parts. This defines a mapping γ from the set of normalised binary Turing machines into $\mathbf{N}$, where

$$\gamma(\mathcal{M}) \ \equiv \ \text{code}(N) \cdot \text{code}(/) \cdot \text{code}(\delta(0,0)) \cdot \text{code}(/) \\ \cdot \text{code}(\delta(0,1)) \cdot \text{code}(/) \cdots \text{code}(\delta(N-1,\mathbf{B})).$$

For a given nonnegative integer ν we can decide whether or not ν belongs to range(γ); in other words, range(γ) is a recursive set. To make this decision, we first note that 0 is not in the range of γ, since all our encodings have leftmost bit equal to 1; so we may assume that $\nu \geq 1$. Identifying ν with its minimal binary representation (the one with the fewest bits), we check whether the number of bits of ν is a multiple of 5; if it is not, then ν cannot belong to range(γ). If the number of bits of ν is a multiple of 5, we split ν into 5-bit blocks and attempt to decode each of these blocks using the inverse of the map code($\cdot$) defined above. If the attempt succeeds, and if the resulting string has the form (3.2) for some nonnegative integer N, then $\nu = \gamma(\mathcal{M})$ for the normalised binary Turing machine $\mathcal{M} \equiv (\mathcal{Q}, \delta, 0, N)$, where $\mathcal{Q} \equiv \{0, 1, ..., N\}$ and the transition function (3.1) is given by the values $\delta(i,j)$ $(0 \leq i \leq N-1)$ read from (3.2); otherwise, $\nu \notin$ range(γ).

From now on, we identify $\mathcal{N}$ with range(γ) whenever it is convenient to do so. Accordingly, we say that a partial function $\varphi : \mathbf{N} \to \mathcal{N}$ (respectively $\psi : \mathcal{N} \to \mathbf{N}$) is **computable** if the partial function $\gamma \circ \varphi : \mathbf{N} \to \mathbf{N}$ (respectively $\psi \circ \gamma^{-1} : \mathbf{N} \to \mathbf{N}$) is computable. In line with the definition on page 35, a total computable function $f : \mathbf{N} \to \mathcal{N}$ is also called an *effective enumeration* of the subset range(f) of $\mathcal{N}$.

(3.6) Theorem.   *There exists a one-one effective enumeration of $\mathcal{N}$ with computable inverse.*

Proof.   By Exercise (3.5.1), there is a strictly increasing total computable function f from $\mathbf{N}$ onto the (clearly infinite) recursive set range(γ). The composite function $\gamma^{-1} \circ f$ is a one-one effective enumeration of $\mathcal{N}$. Moreover,

$$\varphi(n) \ = \ f^{-1}(n) \quad \text{if } n \in \text{ range}(\gamma), \\ = \ \text{undefined} \quad \text{if } n \notin \text{ range}(\gamma)$$

defines a computable partial function $\varphi : \mathbf{N} \to \mathbf{N}$, so the inverse $\varphi \circ \gamma$ of $\gamma^{-1} \circ f$ is a total computable function from $\mathcal{N}$ onto $\mathbf{N}$.   □

In the remainder of this book, $n \mapsto \mathcal{M}_n$ will denote a fixed one-one total computable function from $\mathbf{N}$ onto $\mathcal{N}$ with computable inverse; thus

$$\mathcal{M}_0, \mathcal{M}_1, \mathcal{M}_2, \ldots$$

is an effective enumeration of $\mathcal{N}$. The inverse of this mapping corresponds to an algorithm which, applied to any given normalised binary Turing machine $\mathcal{M}$, produces a unique natural number ν, called the **index** of $\mathcal{M}$, such that $\mathcal{M} = \mathcal{M}_\nu$.

By renaming the states, we can turn any given binary Turing machine into a normalised one. Denoting by $\varphi_i^{(n)}$ the partial function from $\mathbf{N}^n$ to $\mathbf{N}$ computed by $\mathcal{M}_i$, and invoking the Church-Markov-Turing thesis, we therefore see that

$$\varphi_0^{(n)}, \varphi_1^{(n)}, \varphi_2^{(n)}, \ldots \tag{3.3}$$

is an enumeration, which we call the **canonical enumeration**, of the set of all computable partial functions from $\mathbf{N}^n$ to $\mathbf{N}$. For convenience, we usually denote $\varphi_i^{(1)}$ by φ_i.

The natural number ν is known as an **index** of $\varphi_\nu^{(n)}$. Note that to each Turing machine computable partial function $\varphi : \mathbf{N}^n \to \mathbf{N}$ there correspond infinitely many distinct indices i such that $\varphi = \varphi_i^{(n)}$. For, given a normalised binary Turing machine $\mathcal{M}$ that computes φ, we can construct, as follows, distinct normalised binary Turing machines $\mathcal{T}_0 \equiv \mathcal{M}$, $\mathcal{T}_1, \mathcal{T}_2, \ldots$, each of which computes φ : if the halt state of $\mathcal{T}_i$ is m, rename m as $m+1$, adapt the state transition function of $\mathcal{T}_i$ accordingly, and adjoin to $\mathcal{T}_i$ a new state m that cannot be entered from any other state; the resulting normalised binary Turing machine is $\mathcal{T}_{i+1}$.

The canonical enumeration (3.3) is effective in the informal sense that for each $n \geq 1$ there is an algorithm which, applied to an input pair $(i, u) \in \mathbf{N} \times \mathbf{N}^n$, computes $\varphi_i^{(n)}(u)$: simply run $\mathcal{M}_i$ on the input $u_1 0 u_2 0 \ldots 0 u_n$, where $u \equiv (u_1, u_2, \ldots, u_n)$. Thus the enumeration (3.3) should be the prime example of any satisfactory formal notion of an effective enumeration of a set of computable partial functions.

Let $i \mapsto \theta_i$ be an enumeration of a set $\mathcal{S}$ of computable partial functions from $\mathbf{N}^n$ into $\mathbf{N}$; we say that this is an **effective enumeration** of $\mathcal{S}$ if there exists a total computable function $f : \mathbf{N} \to \mathbf{N}$ such that $\theta_i = \varphi_{f(i)}^{(n)}$ for each i. Taking $f(i) \equiv i$, we see immediately that the canonical enumeration is, indeed, effective in this formal sense.

(3.7) Exercises

.1 Construct the encoding of the normalised binary Turing machine described in Figure 11.

.2 In each case describe the Turing machine, if there is one, of which the given binary number is the encoding according to the scheme preceding Theorem (3.6).

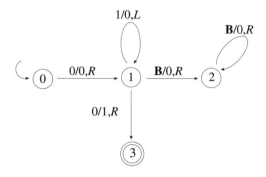

FIGURE 11. The Turing machine in Exercise (3.7.1).

(a)   10011 11111 10001 11111 10001 11111 11101 11111
      11011 11111 10001

(b)   10011 11111 10001 11111 11010 11111 11101 11111
      10010 11111 10001 11111 11101 11111 11011 11111
      10011 11111 11010 11111 11100 11111 11011 11111
      11011 11111 10010 11111 10000 11111 11101 11111
      10011 11111 11010 11111 11100 11111 11011

(c)   10011 11111 10001 11111 10000 11111 11101 11111
      11011 11111 10000 11111 10001 11111 11100

.3   Let S be a recursively enumerable subset of $\mathbf{N}$, and $i \in \mathbf{N}$. Prove
that the partial function $\varphi : \mathbf{N} \to \mathbf{N}$ defined by

$$\varphi(n) \quad = \quad \varphi_i(n) \qquad \text{if } n \in S,$$
$$= \quad \text{undefined} \quad \text{otherwise}$$

is computable.

An interesting application of the Church-Markov-Turing thesis yields
the existence, for each positive integer n, of a **universal Turing machine
for n-ary computable partial functions**: that is, a Turing machine $\mathcal{U}_n$
which, given an input $i0k_10k_20\ldots0k_n$ with $i, k_1, \ldots, k_n \in \mathbf{N}$, computes the
output $\varphi_i^{(n)}(k_1, \ldots, k_n)$. The existence of $\mathcal{U}_n$ follows from the computability
of the partial function $\Phi : \mathbf{N}^{n+1} \to \mathbf{N}$ defined by

$$\Phi(i, k_1, k_2, \ldots, k_n) \equiv \varphi_i^{(n)}(k_1, \ldots, k_n).$$

Of course, using the Church-Markov-Turing thesis to establish the existence
of $\mathcal{U}_n$ is a great deal simpler than writing down the state transition table

or the state diagram of $\mathcal{U}_n$. Details of the construction of $\mathcal{U}_1$ can be found in Section (6.5) of [34] or the Appendix to [27].

We frequently identify the universal Turing machine $\mathcal{U}_n$ with the partial function

$$(i, k_1, k_2, \ldots, k_n) \mapsto \varphi_i^{(n)}(k_1, \ldots, k_n)$$

that it computes.

Given a computable partial function $\theta : \mathbf{N} \to \mathbf{N}$, for future reference we define

$$\varphi_{\theta(i)}^{(n)} \equiv \mathcal{U}_n(\theta(i), \cdot) \quad (i \in \mathbf{N}).$$

This definition accords with our intuition that the partial function $\varphi_{\theta(i)}^{(n)}$ should be everywhere undefined if $i \notin \operatorname{domain}(\theta)$.

Another consequence of the Church-Markov-Turing thesis is that there exists a total function $s : \mathbf{N}^2 \to \mathbf{N}$ such that for all i, j in $\mathbf{N}$,

$$\varphi_i^{(2)}(\cdot, j) = \varphi_{s(i,j)}.$$

Indeed, we can take $s(i, j)$ to be the smallest index of the computable partial function $\varphi_i^{(2)}(\cdot, j) : \mathbf{N} \to \mathbf{N}$. However, this choice does not provide a *computable* function s.[1] The following theorem, a cornerstone of computability theory, shows that we can arrange for s to be a computable function from $\mathbf{N}^2$ to $\mathbf{N}$. Note that, in this theorem and elsewhere, $\mathbf{N}^n \times \mathbf{N}^m$ is identified with $\mathbf{N}^{m+n}$ via the mapping

$$((x_1, \ldots, x_n), (y_1, \ldots, y_m)) \mapsto (x_1, \ldots, x_n, y_1, \ldots, y_m).$$

(3.8) **The s-m-n theorem.**[2]   *For each pair (m, n) of positive integers there exists a total computable function $s : \mathbf{N} \times \mathbf{N}^m \to \mathbf{N}$ such that*

$$\varphi_{s(i,v)}^{(n)} = \varphi_i^{(m+n)}(\cdot, v) \quad (i \in \mathbf{N}, \ v \in \mathbf{N}^m).$$

In other words, for all $i \in \mathbf{N}$ and $v \in \mathbf{N}^m$,

$$\operatorname{domain}(\varphi_{s(i,v)}^{(n)}) = \{u \in \mathbf{N}^n : (u, v) \in \operatorname{domain}(\varphi_i^{(m+n)})\}$$

and

$$\varphi_{s(i,v)}^{(n)}(u) = \varphi_i^{(m+n)}(u, v)$$

whenever either side of this equation is defined.

---

[1]You are asked to prove this in Exercise (5.7.3).

[2]The unimaginative name of this theorem, which is also known as the **parametrisation theorem**, originates with Kleene's notation s_n^m for the function s.

Proof.   We sketch the proof for the case $m = n = 1$. Given $i, j \in \mathbf{N}$, we first construct a binary Turing machine module that, on the input $\lceil k \rceil$, where $k \in \mathbf{N}$, completes a computation with $\lceil k \rceil 0 \lceil j \rceil$ as the output; we can easily arrange that this module $\mathcal{T}(j)$ is defined uniquely by j and has no states in common with $\mathcal{M}_i$. We now append $\mathcal{M}_i$ to the end of $\mathcal{T}(j)$, replacing the halt state of $\mathcal{T}(j)$ by the start state of $\mathcal{M}_i$. The index $s(i, j)$ of the resulting Turing machine can be computed uniquely from i and j. For all $k \in \mathbf{N}$ we clearly have

$$\varphi_{s(i,j)}(k) = \varphi_i^{(2)}(k, j),$$

where the expressions on the left and right of this equation are defined if and only if $(k, j) \in \mathrm{domain}(\varphi_i^{(2)})$.   □

(3.9) Corollary.   *If* $\Phi : \mathbf{N}^2 \to \mathbf{N}$ *is a computable partial function, then there exists a total computable function* $f : \mathbf{N} \to \mathbf{N}$ *such that* $\varphi_{f(i)} = \Phi(i, \cdot)$ *for each* $i \in \mathbf{N}$.

Proof.   By the s-m-n theorem with $m = n = 1$, there exists a total computable function $s : \mathbf{N}^2 \to \mathbf{N}$ such that $\varphi_{s(k,i)} = \varphi_k^{(2)}(\cdot, i)$ for all k and i. Let ν be an index of the computable partial function $(j, i) \mapsto \Phi(i, j)$ on $\mathbf{N}^2$, and set $f(i) \equiv s(\nu, i)$ for each i. Then

$$\varphi_{f(i)} = \varphi_{s(\nu,i)} = \varphi_\nu^{(2)}(\cdot, i) = \Phi(i, \cdot).   □$$

As we shall see in the remaining chapters, the s-m-n theorem is one of the most useful and important tools of computability theory. The first of the next set of exercises gives some idea of how it is applied.

(3.10) Exercises

.1   Prove that there exists a total computable function $F : \mathbf{N}^2 \to \mathbf{N}$ such that $\varphi_{F(i,j)} = \varphi_i \circ \varphi_j$ for all $i, j \in \mathbf{N}$. (This result tells us not only that the composite of two computable partial functions from $\mathbf{N}$ to $\mathbf{N}$ is computable but also how to compute the index of a Turing machine that computes that composite.)

.2*   Prove that for each positive integer n there exists a one-one total computable function c from $\mathbf{N}^n$ onto $\mathbf{N}$ with the following property: if $\psi_i^{(n)} \equiv \varphi_i \circ c$ for each i, then

$$\psi_0^{(n)}, \psi_1^{(n)}, \psi_2^{(n)}, \dots$$

is an effective enumeration of of the set of all Turing machine computable partial functions from $\mathbf{N}^n$ to $\mathbf{N}$.

An enumeration $\psi_0, \psi_1, \ldots$ of the set of all computable partial functions from $\mathbf{N}$ to $\mathbf{N}$ is called an **acceptable programming system** if it has the following properties[3]:

- *s-m-n* **property**: for each computable partial function $\Phi : \mathbf{N}^2 \to \mathbf{N}$ there exists a total computable function $f : \mathbf{N} \to \mathbf{N}$ such that $\psi_{f(i)} = \Phi(i, \cdot)$ for each i.

- **universal property**: the partial function $(i, j) \to \psi_i(j)$ on $\mathbf{N}^2$ is computable.

Theorem (3.8) and the work on page 42 together show that the canonical enumeration $\varphi_0, \varphi_1, \varphi_2, \ldots$ is an acceptable programming system.

Exercises (3.11)

.1   Prove that if $\psi_0, \psi_1, \ldots$ and $\psi'_0, \psi'_1, \ldots$ are acceptable programming systems, then there exists a total computable function $f : \mathbf{N} \to \mathbf{N}$ such that $\psi_n = \psi'_{f(n)}$ for each n. (According to **Rogers' isomorphism theorem**, which we shall not prove here, we can further arrange for the function f to be one-one; see (3.4.7) of [23].)

.2   Prove that an enumeration $\psi_0, \psi_1, \ldots$ of the set of all computable partial functions from $\mathbf{N}$ to $\mathbf{N}$ is an acceptable programming system if and only if there exist total computable functions $f : \mathbf{N} \to \mathbf{N}$ and $g : \mathbf{N} \to \mathbf{N}$ such that $\psi_n = \varphi_{f(n)}$ and $\varphi_n = \psi_{g(n)}$ for each n.

Acceptable programming systems can be used as the basis for a more abstract development of computability theory than ours; see, for example, [23]. We shall only refer to acceptable programming systems once more, in Chapter 5, where they are used to show that a certain consequence of the recursion theorem cannot be improved upon.

---

[3]This is one of several equivalent definitions of an *acceptable programming system*; see pages 94-97 of [23].

4

Computable Numbers and Functions

We begin this chapter by studying in some detail a proof of the fundamental result of computability theory: the undecidability of the halting problem. This will lead us into a discussion of computable real numbers, d-ary expansions, and the elements of computable analysis. You are encouraged to limber up by trying the following exercises.

(4.1) Exercises

.1 Design a Turing machine $\mathcal{M}$ with input alphabet $\{0\}$ and the following property: starting on the left of its tape, $\mathcal{M}$ scans the tape for the first instance of 0; if there is one, $\mathcal{M}$ deletes it and halts with the read/write head on the left; otherwise, $\mathcal{M}$ does not halt.

.2 Prove that there is no Turing machine $\mathcal{M}$ with tape alphabet $\{0, \mathbf{B}\}$ and the following property: when $\mathcal{M}$ is in its start state, with the read/write head scanning the leftmost cell, it moves along the tape looking for all instances of 0; if there is a finite positive number of zeroes on the tape, $\mathcal{M}$ deletes them all and halts with the read/write head on the left; otherwise, $\mathcal{M}$ does not halt.

Computability theory (which, for reasons that will become apparent, might better be called noncomputability theory) deals with such questions as the following:

The Halting Problem: *Is there an algorithm for deciding whether or not a given Turing machine halts on a given input word?*

The Equivalence Problem: *Is there an algorithm for deciding whether or not two given Turing machines compute the same partial function from $\mathbf{N}$ to $\mathbf{N}$?*

The Decidability Problem: *Is every recursively enumerable subset of $\mathbf{N}$ recursive?*

Each of these is an example of a **decision problem**. In such a problem, we consider a property $P(x)$ applicable to elements x of a given set S, and we

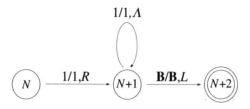

FIGURE 12. The Turing machine module for the proof of Theorem (4.2).

seek an algorithm that will decide, for any given x in S, whether or not $P(x)$ holds; equivalently, we seek a total computable function $f : S \to \{0, 1\}$ such that $f(x) = 1$ if and only if $P(x)$ holds. If there is such an algorithm, we say that the problem, or the property $P(x)$, is **decidable;** otherwise, it is **undecidable.** When $S = \mathbf{N}$, the property $P(x)$ is decidable if and only if $\{x \in \mathbf{N} : P(x)\}$ is a recursive set.

What makes the above questions interesting is the fact that, under the Church-Markov-Turing thesis, the answer in each case is *no.* The basic fact about noncomputability is the undecidability of the halting problem, with which we begin.

(4.2) Theorem. *The total function $f : \mathbf{N} \to \{0, 1\}$ defined by*

$$\begin{aligned} f(n) &= 1 &&\text{if } n \in \text{domain}(\varphi_n), \\ &= 0 &&\text{otherwise} \end{aligned}$$

is not computable.

Proof. Assume that f is computable. Then by the Church-Markov-Turing thesis, f is computable by some normalised binary Turing machine T. Now replace the halt state N of T by the Turing machine module described in Figure 12. Let $\varphi : \mathbf{N} \to \mathbf{N}$ be the partial function computed by the resulting normalised binary Turing machine $\mathcal{M}$. It is easy to see that $\varphi(n)$ is defined if and only if $\ulcorner f(n) \urcorner = 1$; that is, if and only if $f(n) = 0$. Now let ν be the index of $\mathcal{M}$, so that $\varphi = \varphi_\nu$. Then

$$\begin{aligned} \nu \in \text{domain}(\varphi) &\Leftrightarrow f(\nu) = 0 \\ &\Leftrightarrow \varphi(\nu) \text{ is undefined.} \end{aligned}$$

This contradiction completes the proof.   □

It follows from Theorem (4.2) that the halting problem is undecidable.

It is convenient to give here the standard notation for two sets of prime importance in computability theory:

$$K \equiv \{n \in \mathbf{N} : n \in \text{domain}(\varphi_n)\}$$

and its complement

$$\bar{K} \equiv \{n \in \mathbf{N} : n \notin \mathrm{domain}(\varphi_n)\}.$$

Theorem (4.2) immediately leads to the following important result.

(4.3) Corollary. K *is not a recursive set.*   □

This corollary enables us to answer, negatively, the question: *Can every computable partial function from* $\mathbf{N}$ *to* $\mathbf{N}$ *be extended to a total computable function?*

(4.4) Proposition.   *There is a computable partial function, with domain K, that cannot be extended to a total computable function.*

Proof.   Define a computable partial function φ, with domain K, as follows:

$$\begin{aligned} \varphi(n) \quad &= \quad k \qquad\qquad \text{if } \mathcal{M}_n \text{ halts in } k \text{ steps on the input } n, \\ &= \quad \text{undefined} \quad \text{otherwise.} \end{aligned}$$

Suppose there is a total computable function $f : \mathbf{N} \to \mathbf{N}$ such that $f(n) = \varphi(n)$ for each n in K, and consider any $n \in \mathbf{N}$. By our choice of f, if $\varphi_n(n)$ is defined, then it is computed by $\mathcal{M}_n$ in $f(n)$ steps. By running $\mathcal{M}_n$ on the input n and observing whether it halts and has computed $\varphi_n(n)$ in exactly $f(n)$ steps, we can therefore decide whether or not n belongs to K. Hence K is recursive. This contradicts Corollary (4.3).   □

(4.5) Exercises

.1   Prove that K is recursively enumerable. (*Hint*: Use step counting.)

.2   Prove that $\bar{K}$ is not recursively enumerable. (*Hint*: Assume that $\bar{K}$ is recursively enumerable and use Theorem (3.3).)

.3   Prove that a subset S of $\mathbf{N}$ is recursive if and only if both S and $\mathbf{N}\backslash S$ are recursively enumerable. Use this result to give another proof that K is not recursive.

To clarify the proof of Theorem (4.2), let us look more closely at the behaviour of its putative Turing machine $\mathcal{M}$. Consider the following diagram, in which the unparenthesised arrow in position (i, j) (at the intersection of row i and column j) represents the behaviour of the Turing machine $\mathcal{M}_i$ on the (unary form of) the input j: a downward directed arrow says that $\mathcal{M}_i$ halts, and an upward directed arrow says that $\mathcal{M}_i$ fails to halt, when given the input j.

	0	1	2	$\cdots$	n	$\cdots$
$\mathcal{M}_0$	$\uparrow$ ($\downarrow$)	$\uparrow$	$\downarrow$		$\uparrow$	
$\mathcal{M}_1$	$\downarrow$	$\downarrow$ ($\uparrow$)	$\uparrow$		$\downarrow$	
$\mathcal{M}_2$	$\downarrow$	$\downarrow$	$\downarrow$ ($\uparrow$)		$\uparrow$	
$\vdots$						
$\mathcal{M}_n$	$\downarrow$	$\uparrow$	$\uparrow$		$\uparrow$ ($\downarrow$)	
$\vdots$						

The parenthesised arrows indicate the behaviour of $\mathcal{M}$; ($\downarrow$) indicates that $\mathcal{M}$ halts, and ($\uparrow$) that $\mathcal{M}$ fails to halt; thus ($\uparrow$) at position $(2,2)$ indicates that $\mathcal{M}$ fails to halt on the input 2. This example from the diagram illustrates a general feature of the construction of our Turing machine $\mathcal{M}$: *when given the input n, $\mathcal{M}_n$ and $\mathcal{M}$ have precisely the opposite halting behaviour*; if $\mathcal{M}_n$ halts, then $\mathcal{M}$ fails to halt, and if $\mathcal{M}_n$ fails to halt, then $\mathcal{M}$ halts. It follows that $\mathcal{M}$ differs from each of the Turing machines in the list $\mathcal{M}_0, \mathcal{M}_1, \mathcal{M}_2, \ldots$ As this list contains *all* the normalised binary Turing machines, we conclude that $\mathcal{M}$, and hence the required algorithm for deciding the halting problem, cannot possibly exist.

An argument of the type just described, in which the entries along the top-left-to-bottom-right diagonal of a two-by-two array are manipulated to secure a contradiction from certain hypotheses, is called a **diagonal argument**. Diagonal arguments were first used, towards the end of last century, by the German mathematician Georg Cantor to answer the question: *Can we list all the real numbers?* The negative answer to this question is an immediate consequence of **Cantor's Theorem**:

(4.6) Theorem. *If $(a_n)_{n=1}^{\infty}$ is a sequence of real numbers, then in any nondegenerate interval of $\mathbf{R}$ there exists a real number x such that $x \neq a_n$ for each n.*

Proof. Let I_0 be a nondegenerate interval of $\mathbf{R}$; we may assume that I_0 is closed. We construct a sequence $I_0, I_1, I_2, \ldots$ of closed intervals such that for each $n \geq 1$,

(i) I_n is either the left third or the right third of I_{n-1};

(ii) $a_k \notin I_n$ for $k = 1, \ldots, n$.

Assuming that we have constructed $I_0, I_1, \ldots, I_{n-1}$ with the relevant properties, let J_0 be the left third, J_1 the middle third, and J_2 the right third, of I_{n-1}. Either $a_n \notin J_0$, in which case we set $I_n \equiv J_0$; or else $a_n \notin J_2$ and we set $I_n \equiv J_2$. This completes the inductive construction of I_n.

For each k choose a rational number x_k within $|I_k|/6$ of the midpoint of I_k. It follows from (i) that

$$|x_{k+1} - x_k| \le \frac{1}{3}|I_k| + \frac{1}{6}|I_k| + \frac{1}{6}|I_{k+1}| = \frac{5}{9}|I_k| = \frac{5}{9}3^{-k}|I_0|.$$

So for $m > n$,

$$|x_m - x_n| = \left| \sum_{k=n}^{m-1} (x_{k+1} - x_k) \right|$$

$$\le \sum_{k=n}^{m-1} |x_{k+1} - x_k|$$

$$\le \sum_{k=n}^{m-1} \frac{5}{9}3^{-k}|I_0|$$

$$< \frac{5}{9}|I_0| \sum_{k=n}^{\infty} 3^{-k}$$

$$= \frac{5}{18}|I_0|3^{-n+1}$$

$$\rightarrow 0 \text{ as } n \rightarrow \infty.$$

Hence $(x_n)_{n=1}^{\infty}$ is a Cauchy sequence and therefore converges to a limit $x \in \mathbf{R}$. Moreover, for each n, since $x_m \in I_n$ whenever $m \ge n$, and since I_n is closed, we see that $x \in I_n$; whence, by (ii), $x \ne a_n$.   □

The construction of x_n in the above proof need not have been so complicated: it would have sufficed, and would have made the rest of the proof simpler, if we had taken x_n as the midpoint of I_n. We could also have followed Cantor's original approach, which uses decimal expansions. However, by constructing x_n as a *rational* number we have made it possible to use the above proof *mutatis mutandis* to establish a related result about sequences of computable real numbers; see Exercise (4.11.2). Why we chose not to use decimal expansions will be made clear later in this chapter.

A diagram may help to clarify the argument used in our proof of Cantor's Theorem:

	1	2	3	...	n	...
a_1	$R(L)$					
a_2		$L(R)$				
a_3			$L(R)$			
$\vdots$				$\ddots$		
a_n					$R(L)$	
$\vdots$						$\ddots$

In this illustration, if the unparenthesised symbol in position (n, n) is R, then a_n is not in the left third of I_{n-1}; if it is L, then a_n is not in the right third of I_{n-1}. Since we are only interested in the diagonal entries of this array, these are the only ones we have given. The parenthesised symbol in position (n, n) is L if x_n is in the left third of I_{n-1}, and R if x_n is in the right third of I_{n-1}. Again we see the fingerprint of a diagonal argument: the construction of x_n in the proof of Theorem (4.6) is such that x_n is in a third of the interval I_{n-1} that does not contain a_n.

(4.7) Exercises

.1   Construct a mapping of the set of subsets of $\mathbf{N}$ onto the closed interval $[0,1]$. (*Hint*: Consider binary expansions.) Hence show that the set of all subsets of $\mathbf{N}$ is uncountable.

.2   Use the Church-Markov-Turing thesis to prove that the set of all recursively enumerable subsets of $\mathbf{N}$ is countable. Use this fact and Exercise (4.7.1) to give another proof that there exist subsets of $\mathbf{N}$ that are not recursively enumerable. (cf. Exercise (4.5.2)).

What connection, other than the methodological one, exists between the halting problem and Cantor's Theorem that the real numbers cannot be listed? To answer this question, we discuss computable partial functions with values in $\mathbf{Q}$, and define the notion of a computable real number.

By following the arrows through the diagram at the end of this paragraph (overleaf), we obtain an algorithmic enumeration of $\mathbf{Q}$. Moreover, we can remove repetitions from this list to obtain an algorithmic *one-one* enumeration q of $\mathbf{Q}$. By a **computable partial function from $\mathbf{N}^n$ to $\mathbf{Q}$** we mean a partial function $\psi : \mathbf{N}^n \to \mathbf{Q}$ such that the partial function $q^{-1} \circ \psi : \mathbf{N}^n \to \mathbf{N}$ is computable. Hence, by the Church-Markov-Turing thesis,

$$q \circ \varphi_0^{(n)}, \ q \circ \varphi_1^{(n)}, \ q \circ \varphi_2^{(n)}, \ldots$$

is an algorithmic enumeration of the set of computable partial functions from $\mathbf{N}^n$ to $\mathbf{Q}$.

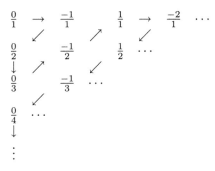

By abuse of language and notation, when the context makes it convenient to do so, we shall identify $\varphi_i^{(n)}$ with $q \circ \varphi_i^{(n)}$ (and φ_i with $q \circ \varphi_\iota$) and refer to $\varphi_i^{(n)}$ as a computable partial function from $\mathbf{N}^n$ to $\mathbf{Q}$.

In elementary analysis courses we learn that every real number is the limit of a sequence of rational numbers; in other words, to each real number x there corresponds a total function $s : \mathbf{N} \to \mathbf{Q}$ such that $|x - s(n)| \to 0$ as $n \to \infty$. We say that x is a **computable real number** if there is a total computable function $s : \mathbf{N} \to \mathbf{Q}$, called a **computable real number generator**, such that $|x - s(n)| \le 2^{-n}$ for each n; otherwise, x is a **noncomputable real number**.

We denote by $\mathbf{R}_c$ the set of computable real numbers.

(4.8) Exercises

.1* Prove that (i) rational numbers, (ii) square roots of positive integers, and (iii) π are computable. (*Hint for* (iii): use a series expansion from calculus.)

.2 Prove that if x is a computable real number, then so is e^x.

.3 Prove that the sum, difference, and product of two computable real numbers are computable.

.4 Prove that if $x \ne 0$ is a computable real number, then $1/x$ is computable.

.5* Prove that if $\varphi_i : \mathbf{N} \to \{0, 1\}$ is total, then

$$\sum_{n=0}^{\infty} (-1)^n 2^{-n} \varphi_i(n)$$

is a computable real number.

.6   Let φ, ψ be computable partial functions from $\mathbf{N}$ to $\mathbf{Q}$. Describe an algorithm which, applied to any $n \in \text{domain}(\varphi) \cap \text{domain}(\psi)$, decides whether $\varphi(n) = \psi(n)$ or $\varphi(n) \neq \psi(n)$.

.7   Let s be a computable real number generator converging to an irrational computable real number x, and let f be a total computable function from $\mathbf{N}$ to $\mathbf{Q}$. Prove that

(i)   for each n there exists k such that $|s(k) - f(n)| > 2^{-k}$;

(ii)   if $s(k) - f(n) > 2^{-k}$, then $x > f(n)$;

(iii)   if $s(k) - f(n) < 2^{-k}$, then $x < f(n)$.

Describe an algorithm which, for each n, decides whether $x > f(n)$ or $x < f(n)$.

The foregoing set of exercises shows that there is a plentiful supply of computable real numbers. What about noncomputable numbers?

(4.9) Theorem.   *There exist noncomputable real numbers; in fact, each nondegenerate interval of* $\mathbf{R}$ *contains noncomputable real numbers.*

Proof.   Let S be the set of those $i \in \mathbf{N}$ such that φ_i is total and such that $|x - q \circ \varphi_i(n)| \leq 2^{-n}$ for some computable real number x and for all n. Clearly, there is an enumeration $i_0, i_1, i_2, \ldots$ of S. Then

$$\lim_{n \to \infty} q \circ \varphi_{i_0}(n), \ \lim_{n \to \infty} q \circ \varphi_{i_1}(n), \ \lim_{n \to \infty} q \circ \varphi_{i_2}(n), \ldots$$

is an enumeration of the set of all computable real numbers. The required conclusion follows immediately from Cantor's Theorem (4.6).   □

In principle, the procedure used in the proof of Cantor's Theorem would enable us to construct an explicit example of a noncomputable real number. In practice, such a number is more easily constructed using the undecidability of the halting problem. First we need a lemma.

(4.10) Lemma.   *There exists a total computable function* $s : \mathbf{N} \to \mathbf{N}$ *such that if* $f : \mathbf{N} \to \{0, 2\}$ *is a total function, and* φ_i *is a computable real number generator converging to* $\sum_{n=0}^{\infty} f(n)3^{-n}$, *then* $f = \varphi_{s(i)}$ *(so that, in particular,* f *is computable).*

Proof.   To begin with, consider a total function $f : \mathbf{N} \to \{0, 2\}$ such that

$$x = \sum_{n=0}^{\infty} f(n)3^{-n}$$

is computable. If φ_i is a computable real number generator converging to x, then for each n,

$$|x - \varphi_i(2n+2)| \le 2^{-2n-2} < 3^{-n}/2.$$

If $f(N) = 0$, then

$$\sum_{n=0}^{N-1} f(n)3^{-n} \le x \le \sum_{n=0}^{N-1} f(n)3^{-n} + 3^{-N},$$

so

$$\varphi_i(2N+2) < \sum_{n=0}^{N-1} f(n)3^{-n} + 3^{-N+1}/2.$$

Similarly, if $f(N) = 2$, then

$$\varphi_i(2N+2) > \sum_{n=0}^{N-1} f(n)3^{-n} + 3^{-N+1}/2.$$

These observations motivate the details of the rest of the proof.

The core of the proof is the definition of a computable partial function $\Psi : \mathbf{N}^2 \to \{0,2\}$ such that for each i, if φ_i is a computable real number generator converging to a computable real number with a ternary expansion $\sum_{n=0}^{\infty} d_n 3^{-n}$ where each d_n belongs to $\{0,2\}$, then $\Psi(i,n) = d_n$. To this end, given $i \in \mathbf{N}$, define $\Psi(i,0) \equiv 0$. Having computed $\Psi(i,0), \ldots, \Psi(i,N-1)$, we define $\Psi(i,N)$ by

$$
\begin{aligned}
\Psi(i,N) &= 0 && \text{if } \varphi_i(2N+2) < \sum_{n=0}^{N-1} \Psi(i,n)3^{-n} + 3^{-N+1}/2, \\
&= 2 && \text{if } \varphi_i(2N+2) > \sum_{n=0}^{N-1} \Psi(i,n)3^{-n} + 3^{-N+1}/2, \\
&= \text{undefined} && \text{otherwise.}
\end{aligned}
$$

Ψ is computable in view of Exercise (4.8.6). By the *s-m-n* theorem, there exists a total computable function $t : \mathbf{N} \to \mathbf{N}$ such that $\varphi_{t(i)} = \Psi(i,\cdot)$ for each i. With x, f, and φ_i as in the observations at the beginning of the proof, we see that $\varphi_{t(i)}$ is total and that $\varphi_{t(i)}(n) = f(n)$ for each n. Hence

$$x = \sum_{n=0}^{\infty} \varphi_{t(i)}(n)3^{-n}.$$

This completes the proof. $\square$

We are now in a position to show that the real number

$$x \equiv \sum_{n=0}^{\infty} \chi_K(n)3^{-n},$$

which is well defined since the series on the right converges by comparison with $\sum_{n=0}^{\infty} 3^{-n}$, is noncomputable. Indeed, if x were computable, then, by Lemma (4.10) applied to $2x$, the total function $2\chi_K : \mathbf{N} \rightarrow \{0,2\}$, and therefore χ_K, would be computable, which would contradict Corollary (4.3).

In order to understand this number x a little better, consider how we might try to calculate its ternary expansion. To produce the nth ternary digit of x, we give the Turing machine $\mathcal{M}_n$ the input n and see what happens. If we are lucky, $\mathcal{M}_n$ will halt on the input n fairly quickly, and we will be able to write down 1 as the nth ternary digit of x; if, however, we are unlucky, $\mathcal{M}_n$ may carry on executing longer than the life of the universe, and we will have no way of knowing whether it carries on for ever or halts some time after we are no longer interested in the affair.

It is important to realise that the noncomputability of x is not a matter of inadequate computing power, precision, or memory: x is noncomputable *in principle* as well as in practice.

(4.11) Exercises

.1   Give a diagonal argument to prove Cantor's Theorem using decimal expansions. (Reduce to the case where each term of the given sequence of real numbers is between 0 and 1. It will help if you then prove the following lemma: If $a \equiv a_0 \cdot a_1 a_2 \ldots$ and $b \equiv b_0 \cdot b_1 b_2 \ldots$ are decimal expansions of real numbers such that a_n and b_n differ by at least 2 *modulo* 10 for some n, then $a \neq b$.)

.2   Let f be a total computable function from $\mathbf{N}$ to $\mathbf{Q}$ such that $\varphi_{f(n)}$ is a computable real number generator for each n. Prove that there exists ν such that φ_ν is a computable real number generator and such that

$$\lim_{k \to \infty} \varphi_\nu(k) \neq \lim_{k \to \infty} \varphi_{f(n)}(k)$$

for all n.

.3   Give at least two proofs that an increasing binary sequence is computable.

We say that a partial function $\Theta : \mathbf{N} \rightarrow \mathbf{R}_c$ is **computable** if there exists a computable partial function $\theta : \mathbf{N} \rightarrow \mathbf{N}$ such that if $i \in \text{domain}(\Theta)$, then $i \in \text{domain}(\theta)$ and $\varphi_{\theta(i)}$ is a computable real number generator converging to $\Theta(i)$. (Note that $\theta(i)$ may be defined even if $\varphi_{\theta(i)}$ is not a computable real number generator.) Similar definitions apply to notions such as that of a computable partial function from $\mathbf{Q}$ to $\mathbf{R}_c$ and that of a computable partial function from $\mathbf{R}_c$ into $\mathbf{N}$; the formulation of these definitions is left to Exercise (4.14.1).

We now see from Exercise (4.11.2) that if (a_n) is a computable sequence of computable real numbers, then there is a computable real number a such that $a \neq a_n$ for all n. Thus although, as we observed in the proof of Theorem (4.9), the set of computable real numbers can be enumerated, there is no *effective* enumeration of that set.

We say that a partial function $\Theta : \mathbf{R}_c \rightarrow \mathbf{R}_c$ is **computable** if there exists a computable partial function $\theta : \mathbf{N} \rightarrow \mathbf{N}$ such that if $\varphi_i : \mathbf{N} \rightarrow \mathbf{Q}$ is a computable real number generator that converges to a limit x in the domain of Θ, then $i \in \mathrm{domain}(\theta)$ and $\varphi_{\theta(i)}$ is a computable real number generator that converges to $\Theta(x)$. Related notions of computable partial function, such as one from $\mathbf{R}_c \times \mathbf{R}_c$ to $\mathbf{R}_c$, are defined analogously in the obvious way.

Naturally, we hope that our definitions of computable real number and computable partial function from $\mathbf{R}_c \times \mathbf{R}_c$ to $\mathbf{R}_c$ will enable us to prove that the elementary arithmetic operations on computable real numbers are performed by computable partial functions. Our approach to computable real numbers, through computable rational approximating sequences rather than computable decimal expansions, was chosen to ensure that this is, indeed, the case.

(4.12) Proposition.   *The total functions* **plus**, **minus**, *and* **times**, *defined on* $\mathbf{R}_c \times \mathbf{R}_c$ *by*

$$\mathbf{plus}(x, y) \quad \equiv \quad x + y,$$
$$\mathbf{minus}(x, y) \quad \equiv \quad x - y,$$
$$\mathbf{times}(x, y) \quad \equiv \quad xy,$$

are computable.

Proof.   Exercise (4.14.2).   □

A little more difficult to prove is

(4.13) Proposition.   *The partial function* $\mathbf{div} : \mathbf{R}_c \times \mathbf{R}_c \rightarrow \mathbf{R}_c$, *defined by*

$$\mathbf{div}(x, y) \equiv x/y \text{ whenever } y \neq 0,$$

is computable.

Proof.   In view of the computability of the function **times** (Proposition (4.12)) and of the composition of computable partial functions from $\mathbf{R}_c$ to $\mathbf{R}_c$ (Exercise (4.14.4)), it will suffice to prove that there exists a total computable function $f : \mathbf{N} \rightarrow \mathbf{N}$ such that if φ_m is a computable real number generator converging to a nonzero computable real number, then $\varphi_{f(m)}$ is a computable real number generator and

$$\lim_{n\to\infty} \varphi_{f(m)}(n) = 1/\lim_{n\to\infty} \varphi_m(n).$$

To this end, for each $m \in \mathbf{N}$ define

$$\psi(m) = \min k \left[k \geq 2 \text{ and } |\varphi_m(k)| > 2^{-k+2} \right].$$

By Exercises (2.6.5) and (2.7.3), ψ is a computable partial function from $\mathbf{N}$ to $\mathbf{N}$. Next, define a computable partial function $\Psi : \mathbf{N}^2 \to \mathbf{N}$ by

$$
\begin{aligned}
\Psi(m,n) \quad &= \quad 1/\varphi_m(2\psi(m) + n - 2) \quad \text{if } \varphi_m(2\psi(m) + n - 2) \text{ is} \\
&\qquad\qquad\qquad\qquad\qquad\qquad \text{defined and nonzero,} \\
&= \quad \text{undefined} \qquad\qquad\quad \text{otherwise.}
\end{aligned}
$$

Applying Corollary (3.9) to Ψ, we can find a total computable function $f : \mathbf{N} \to \mathbf{N}$ such that $\varphi_{f(m)} = \Psi(m, \cdot)$ for each m. Consider any $m \in \mathbf{N}$ such that φ_m is a computable real number generator converging to a nonzero limit x. Since x is nonzero, $\psi(m)$ is defined; also, for all $k \geq \psi(m)$ we have

$$
\begin{aligned}
|\varphi_m(k)| \quad &\geq \quad |\varphi_m(\psi(m))| - |x - \varphi_m(\psi(m))| - |x - \varphi_m(k)| \\
&> \quad 2^{-\psi(m)+2} - 2^{-\psi(m)} - 2^{-k} \\
&\geq \quad 2^{-\psi(m)+2} - 2^{-\psi(m)} - 2^{-\psi(m)} \\
&= \quad 2^{-\psi(m)+1};
\end{aligned}
$$

whence $|x|^{-1} \leq 2^{\psi(m)-1}$. Since $\psi(m) \geq 2$, for each $n \in \mathbf{N}$ we have

$$2\psi(m) + n - 2 \geq \psi(m)$$

and therefore

$$\varphi_m(2\psi(m) + n - 2) > 2^{-\psi(m)+1};$$

whence $\varphi_{f(m)}(n)$ is defined and

$$
\begin{aligned}
&\left| 1/x - \varphi_{f(m)}(n) \right| \\
&= \quad |x|^{-1} |\varphi_m(2\psi(m) + n - 2)|^{-1} |x - \varphi_m(2\psi(m) + n - 2)| \\
&< \quad 2^{\psi(m)-1} 2^{\psi(m)-1} 2^{-2\psi(m)-n+2} \\
&= \quad 2^{-n}.
\end{aligned}
$$

This shows that $\varphi_{f(m)}$, each of whose values is certainly rational, is a computable real number generator converging to $1/x$.

(4.14) Exercises

.1   Formulate the definition of a computable partial function (i) from $\mathbf{R}_c$ to $\mathbf{N}$, (ii) from $\mathbf{Q}$ to $\mathbf{R}_c$, and (iii) from $\mathbf{N} \times \mathbf{R}_c$ to $\mathbf{R}_c$.

.2   Prove Proposition (4.12).

.3   Explain why, in the proof of Proposition (4.13), the partial function $\Psi : \mathbf{N}^2 \to \mathbf{N}$ is computable.

.4   Prove that if $\Theta : \mathbf{R}_c \to \mathbf{R}_c$, $\Psi : \mathbf{N} \to \mathbf{R}_c$, and $\Psi' : \mathbf{R}_c \to \mathbf{R}_c$ are computable partial functions, then so are the composite functions $\Theta \circ \Psi$ and $\Theta \circ \Psi'$.

.5   Prove that if $\Theta : \mathbf{R}_c \times \mathbf{R}_c \to \mathbf{N}$ is a computable partial function, then for each $a \in \mathbf{R}_c$ the partial function $x \mapsto \Theta(x, a)$ from $\mathbf{R}_c$ to $\mathbf{N}$ is computable.

.6   Let $\Theta : \mathbf{N} \to \mathbf{R}_c$ be a partial function, and define a corresponding partial function $\Theta^* : \mathbf{R}_c \to \mathbf{R}_c$ by

$$\begin{aligned} \Theta^*(x) &= \Theta(x) &&\text{if } x \in \mathbf{N}, \\ &= \text{undefined} &&\text{if } x \notin \mathbf{N}. \end{aligned}$$

Prove that Θ^* is computable if and only if Θ is computable. This example reconciles two apparently different notions of a computable partial function from $\mathbf{N}$ into $\mathbf{R}_c$ that arise from the definitions on pages 56-57.

If you have completed Exercise (4.11.1), you may be wondering why we chose not to define computable real numbers in terms of decimal or other expansions. The following definition and results prepare us for the explanation of that choice.

Given an integer $d \geq 2$, we say that a real number x has a **computable d-ary expansion** if there exist $j \in \{0, 1\}$ and a total computable function $f : \mathbf{N} \to \mathbf{N}$ such that $f(n) \in \{0, 1, ..., d - 1\}$ for all $n \geq 1$ and such that $x = (-1)^j \sum_{n=0}^{\infty} f(n)d^{-n}$. It is a simple exercise to prove that if a real number x has a computable d-ary expansion, then x is a computable real number. In fact, we can prove more.

(4.15) Proposition.   *If $d \geq 2$ is an integer, then there exists a total computable function $g : \mathbf{N}^2 \to \mathbf{N}$ with the following property: if*

$$(-1)^j \sum_{k=0}^{\infty} \varphi_m(k)d^{-k}$$

is the d-ary expansion of a real number x, then

$$\left| x - \varphi_{g(j,m)}(n) \right| < d^{-n}$$

for each n (where $\varphi_{g(j,m)}$ is considered as a function from $\mathbf{N}$ to $\mathbf{Q}$). In particular, $\varphi_{g(j,m)}$ is a computable real number generator converging to x.

Proof.   Define a total computable function $F : \mathbf{N}^3 \to \mathbf{N}$ by

$$F(j, m, n) = (-1)^j \sum_{k=0}^{n+1} \varphi_m(k)d^{-k}.$$

By the s-m-n theorem, there exists a total computable function $g : \mathbf{N}^2 \to \mathbf{N}$ such that $\varphi_{g(j,m)} = F(j, m, \cdot)$ for all $j, m \in \mathbf{N}$. If $(-1)^j \sum_{k=0}^{\infty} \varphi_m(k)d^{-k}$ is the d-ary expansion of a real number x, then for each n we have

$$\left| x - \varphi_{g(j,m)}(n) \right| \leq \sum_{k=n+2}^{\infty} \varphi_m(k)d^{-k}$$

$$\leq (d-1) \sum_{k=n+2}^{\infty} d^{-k}$$

$$= d^{-n-1}$$

$$< d^{-n}.$$

The final statement of the theorem follows because $d^{-n} \leq 2^{-n}$.   □

As we shall show subsequently, the following is the best we can hope for by way of a general converse of Proposition (4.15).

(4.16) Proposition.   *If $d \geq 2$ is an integer, then each computable real number x has a computable d-ary expansion.*

Proof.   It is a routine exercise in elementary school arithmetic to show that if x is rational, then it has a computable d-ary expansion: indeed, choosing $j \in \{0, 1\}$, a natural number m, and a positive integer n such that $x = (-1)^j m/n$, and working relative to the base d, we carry out the long division of m by n and then multiply by $(-1)^j$. So we may assume that x is a computable irrational number. By Exercise (4.8.7), we can then decide, for each integer n, whether $x < n$ or $x > n$; so we may assume that $0 < x < 1$. Now suppose that we have found natural numbers $n_0 \equiv 0, ..., n_k$ such that $0 \leq n_j \leq d - 1$ $(1 \leq j \leq k)$, and such that

$$\sum_{j=0}^{k} n_j d^{-j} < x < \sum_{j=0}^{k-1} n_j d^{-j} + (n_k + 1)d^{-k}.$$

Again applying Exercise (4.8.7), we can find a unique $t \in \{0, ..., d-1\}$ such that

$$\sum_{j=0}^{k} n_j d^{-j} + td^{-k-1} < x < \sum_{j=0}^{k} n_j d^{-j} + (t+1)d^{-k-1};$$

we then take n_{k+1} equal to this value t. Setting $f(k) \equiv n_k$ for each k, we have described the inductive construction of a total, clearly computable, function $f : \mathbf{N} \to \mathbf{N}$ such that $f(k) \in \{0, ..., d-1\}$ for each $k \geq 1$ and such that $x = \sum_{k=0}^{\infty} f(k) d^{-k}$. $\square$

(4.17) Exercises

.1 Given an integer $d \geq 2$, prove that there exist total computable functions $r : \mathbf{N} \to \mathbf{N}$, $s : \mathbf{N} \to \{0, 1\}$ such that for each i, $\varphi_{r(i)}$ is a total computable function from $\mathbf{N}$ to $\{0, ..., d-1\}$ and

$$q(i) \equiv (-1)^{s(i)} \sum_{n=0}^{\infty} \varphi_{r(i)}(n) d^{-n},$$

where q is the one-one total computable function from $\mathbf{N}$ onto $\mathbf{Q}$ introduced on page 52.

.2 Given an integer $d \geq 2$, prove that there exists a total computable function $s : \mathbf{N} \to \mathbf{N}$ such that if φ_i is a computable real number generator converging to a positive irrational number $x \in \mathbf{R}_c$, then $\varphi_{s(i)}$ is a total computable function from $\mathbf{N}$ to $\{0, ..., d-1\}$ and

$$x = \sum_{n=0}^{\infty} \varphi_{s(i)}(n) d^{-n}.$$

(*Hint*: Model your proof on that of Proposition (4.16), with reference to Exercise (4.8.7).)

Although each computable real number x has a computable d-ary expansion, in the proof of Proposition (4.16) the algorithm for computing that expansion depends on whether x is rational or irrational. This is unavoidable. For if there were a single algorithm, applicable to each computable real number x, that computed a d-ary expansion of x, then there would be a total computable function $f : \mathbf{R}_c \to \mathbf{N}$ such that for each $x \in \mathbf{R}_c$, $f(x)$ was the integer part of a binary expansion of x; as we now aim to show, no such computable function f exists. It is for this reason that we described Proposition (4.16) as the best possible converse of its predecessor.

(4.18) Lemma.   *There exists a total computable function $F : \mathbf{N}^2 \to \{0, 1\}$ such that*

(i)   *for each m there is at most one n such that $F(m, n) = 1$;*
(ii) *if t is a total computable function from $\mathbf{N}$ to $\{0, 1\}$, then there exist m and k in $\mathbf{N}$ such that $F(m, 2k + t(m)) = 1$.*

Proof.   Set

$$F(m,n) \quad = \quad 1 \quad \text{if } \mathcal{M}_m \text{ halts in } k \text{ steps on the input } m, \text{ and}$$
either $n = 2k$ and $\varphi_m(m) = 0$, or $n = 2k+1$
and $\varphi_m(m) > 0$,

$$= \quad 0 \quad \text{otherwise.}$$

Clearly, $F : \mathbf{N}^2 \to \{0,1\}$ is total and computable, and satisfies (i). Given a total computable function $t : \mathbf{N} \to \{0,1\}$, choose m such that $t = \varphi_m$, and then k such that $\mathcal{M}_m$ halts in k steps on the input m. If $t(m) = 1$, then $\varphi_m(m) = 1$ and therefore $F(m, 2k+1) = 1$; whereas if $t(m) = 0$, then $\varphi_m(m) = 0$ and therefore $F(m, 2k) = 0$. Thus F satisfies (ii). $\square$

A moment's reflection should convince you of the improbability of finding an algorithm which, applied to a computable binary sequence with at most one term equal to 1, shows either that all the even-indexed terms of the sequence are 0 or that all the odd-indexed terms are 0. The next proposition confirms that conviction.

(4.19) Proposition.   *There is no computable partial function $\theta : \mathbf{N} \to \{0,1\}$ such that if $\varphi_i : \mathbf{N} \to \{0,1\}$ is total and $\varphi_i(n) = 1$ for at most one n, then*

(i) $i \in \mathrm{domain}(\theta)$,
(ii) $\theta(i) = 0 \Rightarrow \varphi_i(n) = 0$ *for all even n, and*
(iii) $\theta(i) = 1 \Rightarrow \varphi_i(n) = 0$ *for all odd n.*

Proof.   Let F be as in Lemma (4.18), and, using Corollary (3.9), construct a total computable function $s : \mathbf{N} \to \mathbf{N}$ such that $\varphi_{s(i)} = F(i, \cdot)$ for each i. Suppose there exists a computable partial function $\theta : \mathbf{N} \to \mathbf{N}$ with the properties described in the statement of this proposition. Then $\theta \circ s$ is a total computable function from $\mathbf{N}$ into $\{0,1\}$; whence, by Lemma (4.18), there exist m, k such that $F(m, 2k + \theta(s(m))) = 1$. If $\theta(s(m)) = 0$, then $F(m, 2k) = 1$; also, by the assumed property of θ, $F(m,n) = \varphi_{s(m)}(n) = 0$ for all even n—a contradiction. Similarly, if $\theta(s(m)) = 1$, then $F(m, 2k + 1) = 1$, and $F(m,n) = 0$ for all odd n—again a contradiction. $\square$

(4.20) Lemma.   *There is no computable partial function $\theta : \mathbf{N} \to \{0,1\}$ such that if $\varphi_i : \mathbf{N} \to \{0,1\}$ is total, then*

(i) $i \in \mathrm{domain}(\theta)$, *and*
(ii) $\theta(i)$ *is the integer part of* $1 + \sum_{n=0}^{\infty} (-1)^n 2^{-n} \varphi_i(n)$.

Proof. Suppose such a computable partial function θ exists. Consider any i such that φ_i is a total computable function from $\mathbf{N}$ to $\{0,1\}$, and such that $\varphi_i(n) = 1$ for at most one n. If $\varphi_i(k) = 1$ for an even k, then

$$1 + \sum_{n=0}^{\infty} (-1)^n 2^{-n} \varphi_i(n) = 1 + 2^{-k} > 1$$

and so $\theta(i) = 1$. Thus if $\theta(i) = 0$, then $\varphi_i(k) = 0$ for all even k. Similarly, if $\theta(i) = 1$, then $\varphi_i(k) = 0$ for all odd k. These conclusions contradict Proposition (4.19). $\square$

(4.21) Proposition. *There is no total computable function $F : \mathbf{R}_c \times \mathbf{R}_c \to \mathbf{Q}$ such that for all x, y in $\mathbf{R}_c$, $F(x,y)$ is the integer part of a binary expansion of $x + y$.*

Proof. Suppose such a function F exists. Then, in view of Exercises (4.14.1) and (4.14.5), there exists a computable partial function $\psi : \mathbf{N} \to \mathbf{N}$ such that if φ_k is a real number generator converging to the computable real number x, then $k \in \operatorname{domain}(\psi)$ and $\psi(k) = F(x, 1)$. Define a computable partial function $G : \mathbf{N}^2 \to \mathbf{N}$ by

$$G(i,j) = \sum_{n=0}^{j} (-1)^n 2^{-n} \max\{0, \min\{1, \varphi_i(n)\}\}.$$

By Corollary (3.9), there exists a total computable function $s : \mathbf{N} \to \mathbf{N}$ such that $\varphi_{s(i)} = G(i, \cdot)$ for each i. Let $\theta \equiv \psi \circ s$, and consider any i such that φ_i is a total computable function from $\mathbf{N}$ into $\{0,1\}$. A simple computation shows that $\varphi_{s(i)}$ is a computable real number generator converging to the computable real number $\sum_{n=0}^{\infty} (-1)^n 2^{-n} \varphi_i(n)$; so $i \in \operatorname{domain}(\theta)$ and

$$\theta(i) = \psi(s(i)) = F\left(\sum_{n=0}^{\infty} (-1)^n 2^{-n} \varphi_i(n), 1\right).$$

This contradicts Lemma (4.20). $\square$

It follows from Proposition (4.21) that there is no total computable function $f : \mathbf{R}_c \to \mathbf{N}$ such that for each $x \in \mathbf{R}_c$, $f(x)$ is the integer part of a binary expansion of x. This completes the justification of our claim, on page 61, that the algorithm for computing the binary expansion of $x \in \mathbf{R}_c$ depends on x.

(4.22) Proposition.   *There is no computable partial function* $\sigma : \mathbf{N}^2 \to$ $\mathbf{N}$ *such that if* φ_i, φ_j *are total computable functions from* $\mathbf{N}$ *to* $\{0, 1\}$, *then* $(i, j) \in \text{domain}(\sigma)$ *and*

$$\sum_{n=0}^{\infty} \varphi_{\sigma(i,j)}(n) 2^{-n} = \sum_{n=0}^{\infty} (\varphi_i(n) + \varphi_j(n)) 2^{-n}.$$

Proof.   The idea underlying this proof is quite simple. Given a binary sequence (a_n) with $a_0 = 0$ and with at most one term equal to 1, set $x_0 = y_0 = 0$, and for $n \geq 1$ define

$$
\begin{aligned}
x_n &= 1 & &\text{if } a_k = 0 \text{ for each odd } k \leq n, \\
&= 0 & &\text{otherwise}, \\
y_n &= 0 & &\text{if } a_k = 0 \text{ for each even } k \leq n, \\
&= 1 & &\text{otherwise}.
\end{aligned}
$$

Setting

$$x \equiv \sum_{n=0}^{\infty} x_n 2^{-n},$$

$$y \equiv \sum_{n=0}^{\infty} y_n 2^{-n},$$

we see that $x + y > 1$ if $a_n = 1$ for an even value of n, and $x + y < 1$ if $a_n = 1$ for an odd value of n. So by looking at the integer part of $x + y$ we can tell whether $a_n = 1$ for an even value of n or $a_n = 1$ for an odd value of n.

To make this idea more precise, define computable partial functions $\Theta, \Psi : \mathbf{N}^2 \to \mathbf{N}$ as follows:

$$\Theta(i, 0) = \Psi(i, 0) = 0,$$

and for each $n \geq 1$,

$$
\begin{aligned}
\Theta(i, n) &= 1 & &\text{if } \varphi_i(k) \text{ is defined for all } k \leq n, \text{ and} \\
& & &\varphi_i(k) = 0 \text{ for all odd } k \leq n, \\
&= 0 & &\text{if } \varphi_i(k) \text{ is defined for all } k \leq n, \text{ and} \\
& & &\varphi_i(k) = 1 \text{ for some odd } k \leq n, \\
&= \text{undefined} & &\text{otherwise}.
\end{aligned}
$$

$$
\begin{aligned}
\Psi(i, n) &= 0 & &\text{if } \varphi_i(k) \text{ is defined for all } k \leq n, \text{ and} \\
& & &\varphi_i(k) = 0 \text{ for all even } k \leq n, \\
&= 1 & &\text{if } \varphi_i(k) \text{ is defined for all } k \leq n, \text{ and} \\
& & &\varphi_i(k) = 1 \text{ for some even } k \leq n, \\
&= \text{undefined} & &\text{otherwise}.
\end{aligned}
$$

Note that if φ_i is a total function from $\mathbf{N}$ into $\{0,1\}$, then $\Theta(i,n)$ and $\Psi(i,n)$ are defined for all n. By the s-m-n theorem, there exist total computable functions s, $t : \mathbf{N} \to \mathbf{N}$ such that $\varphi_{s(i)} = \Theta(i,\cdot)$ and $\varphi_{t(i)} = \Psi(i,\cdot)$ for each i. Now suppose there exists a computable partial function $\sigma : \mathbf{N}^2 \to \mathbf{N}$ with the stated properties. Define a computable partial function $\theta : \mathbf{N} \to \{0,1\}$ by

$$\theta(i) \equiv \varphi_{\sigma \circ (s,t)(i)}(0).$$

Consider any index i such that $\varphi_i : \mathbf{N} \to \{0,1\}$ is total and such that $\varphi_i(n) = 1$ for at most one value of n. Let

$$x \equiv \sum_{n=0}^{\infty} \varphi_{s(i)}(n) 2^{-n},$$

$$y \equiv \sum_{n=0}^{\infty} \varphi_{t(i)}(n) 2^{-n}.$$

Then $\theta(i)$ is defined, and

$$x + y = \sum_{n=0}^{\infty} \varphi_{\sigma \circ (s,t)(i)}(n) 2^{-n}.$$

If $\varphi_i(2j) = 1$, then $x = \sum_{n=0}^{\infty} 2^{-n}$, $y = \sum_{n=2j}^{\infty} 2^{-n}$, $x + y > 1$, and so $\theta(i) = 1$; whereas if $\varphi_i(2j+1) = 1$, then $x = \sum_{n=0}^{2j} 2^{-n}$, $y = 0$, $x + y < 1$, and so $\theta(i) = 0$. Thus if $\theta(i) = 0$, then $\varphi_i(n) = 0$ for all even n; and if $\theta(i) = 1$, then $\varphi_i(n) = 0$ for all odd n. This contradicts Proposition (4.19). $\square$

It should now be clear why we chose not to define computable real numbers in terms of binary (or d-ary) expansions: had we done so, we would have had the unsatisfactory situation in which the addition of x and y could not be performed by a computable partial function of the indices of total computable functions giving the binary digits of x and y.

(4.23) Exercises

In the following exercises remember our identification of φ_i with $q \circ \varphi_i$, where $q : \mathbf{N} \to \mathbf{Q}$ is the effective enumeration of $\mathbf{Q}$ introduced on page 52.

.1   Prove that there is no computable partial function $\theta : \mathbf{N} \to \mathbf{N}$ such that if $\varphi_i : \mathbf{N} \to \{0,1\}$ is total, then $i \in \mathrm{domain}(\theta)$,

$$\begin{aligned}
\theta(i) = 0 &\Rightarrow \varphi_i(n) = 0 \text{ for all } n, \text{ and} \\
\theta(i) = 1 &\Rightarrow \text{there exists } n \text{ such that } \varphi_i(n) = 1.
\end{aligned}$$

.2   Prove that there is no computable partial function $\theta : \mathbf{N} \rightarrow \{0,1\}$ such that if φ_i is a computable real number generator, then $i \in$ domain(θ),

$$\theta(i) = 0 \quad \Rightarrow \quad \lim_{n \to \infty} \varphi_i(n) < 0, \text{ and}$$
$$\theta(i) = 1 \quad \Rightarrow \quad \lim_{n \to \infty} \varphi_i(n) \geq 0.$$

.3   Prove that there is no total computable function $f : \mathbf{R}_c \rightarrow \{0,1\}$ such that

$$f(x) = 0 \quad \Rightarrow \quad x = 0, \text{ and}$$
$$f(x) = 1 \quad \Rightarrow \quad x \neq 0.$$

(*Hint*: Consider binary expansions of the form $\sum_{n=0}^{\infty} 2^{-n}\varphi_i(n)$, where φ_i is a total computable function from $\mathbf{N}$ into $\{0,1\}$.) Thus there is no algorithm which, applied to any computable real number x, will decide whether $x = 0$ or $x \neq 0$. Is there an algorithm which, applied to any *rational* number x, will decide whether $x = 0$ or $x \neq 0$?

.4   Prove that there is no algorithm for deciding whether or not a given real number is rational; more precisely, prove that there is no total computable function $f : \mathbf{R}_c \rightarrow \{0,1\}$ such that

$$f(x) = 0 \quad \Rightarrow \quad x \text{ is rational, and}$$
$$f(x) = 1 \quad \Rightarrow \quad x \text{ is irrational.}$$

(*Hint*: Consider real numbers of the form $\sum_{n=0}^{\infty} \varphi_i(n)/n!$, where $\varphi_i : \mathbf{N} \rightarrow \{0,1\}$ is total and increasing.[1])

So far, we have discussed only the most basic properties of the computable real number line $\mathbf{R}_c$, which stands at the entrance to the remarkable world of *computable*, or *recursive, analysis*. We end this chapter by taking a few steps into the interior of that world.

In recursive mathematics we work with effective analogues of the standard notions and properties found in traditional mathematics. For example, when we are dealing with the convergence of sequences, we work with an effective notion of convergence in which the rate of convergence to the limit is expressed by a computable function; to be precise, we say that a sequence (x_n) of real numbers **converges effectively** to a real number x if there exists a total computable function $h : \mathbf{N} \rightarrow \mathbf{N}$ such that $|x_n - x| \leq 2^{-N}$ whenever $n \geq h(N)$.

For the following exercises we define a **computable sequence of computable partial functions from $\mathbf{R}_c$ to $\mathbf{R}_c$** to be a sequence $(f_n)_{n=0}^{\infty}$ of

---

[1]We say that a partial function $\varphi : \mathbf{R}_c \rightarrow \mathbf{R}_c$ is **increasing** if $\varphi(x) \leq \varphi(x')$ whenever $x, x' \in$ domain(φ) and $x \leq x'$. Some authors would describe such a function φ as *nondecreasing*.

such functions with the property that the partial function $(n, x) \mapsto f_n(x)$ from $\mathbf{N} \times \mathbf{R}_c$ to $\mathbf{R}_c$ is computable. Thus $(f_n)_{n=0}^{\infty}$ is a computable sequence if and only if there exists a computable partial function $\theta : \mathbf{N}^2 \to \mathbf{N}$ such that if φ_i is a computable real number generator converging to a point x of domain(f_n), then $(n, i) \in$ domain(θ) and $\varphi_{\theta(n,i)}$ is a computable real number generator converging to $f_n(x)$.

(4.24) Exercises

.1   Prove that the limit of an effectively convergent computable sequence of computable real numbers is a computable real number.

.2   Prove **Specker's Theorem**: *There exists a strictly increasing computable sequence (a_n) of rational numbers in $[0,1]$ that does not converge effectively.* (*Hint*: Let $f : \mathbf{N} \to K$ be an effective enumeration of K, and define $a_n \equiv \sum_{m=0}^{n} 2^{-f(m)-1}$.)

.3   Let (f_n) be a computable sequence of total computable functions from $\mathbf{R}_c$ to $\mathbf{R}_c$, and for each n let $s_n \equiv \sum_{k=0}^{n} f_k$. Prove that (s_n) is a computable sequence of total computable functions from $\mathbf{R}_c$ to $\mathbf{R}_c$.

.4   Let $F, F_0, F_1, \ldots$ be total functions from $\mathbf{R}$ to $\mathbf{R}$. Suppose that there exists a total computable function $h : \mathbf{N} \to \mathbf{N}$ such that

$$|F(x) - F_n(x)| \le 2^{-N}$$

whenever $x \in \mathbf{R}$ and $n \ge h(N)$; in which case we say that the sequence (F_n) **converges effectively and uniformly** to F. Let f, f_n be respectively the restrictions of F, F_n to $\mathbf{R}_c$. Prove that if $(f_n)_{n=0}^{\infty}$ is a computable sequence of total computable functions from $\mathbf{R}_c$ to $\mathbf{R}_c$, then f is a total computable function from $\mathbf{R}_c$ to $\mathbf{R}_c$.

.5   Let $F_0, F_1, \ldots$ be total functions from $\mathbf{R}$ to $\mathbf{R}$, and for each n let f_n be the restriction of F_n to $\mathbf{R}_c$. Suppose that each f_n maps $\mathbf{R}_c$ to $\mathbf{R}_c$, and that $(f_n)_{n=0}^{\infty}$ is a computable sequence of total computable functions from $\mathbf{R}_c$ to $\mathbf{R}_c$. Prove that for each computable sequence $(x_k)_{k=0}^{\infty}$ of computable real numbers there exists a computable double sequence $(r_{n,k})_{n,k=0}^{\infty}$ of rational numbers such that $|f_n(x_k) - r_{n,k}| \le 2^{-k}$ for all n and k.

We shall return to effective convergence at the end of this chapter. In the meantime, we examine the recursive content of the **Heine-Borel Theorem**: every open cover of $[0,1]$ contains a finite subcover.

By an **effective sequence of open intervals in R** we mean a total computable function $f : \mathbf{N} \to \mathbf{R}_c \times \mathbf{R}_c$; informally, we identify the ordered pair

$$(P_1^2 \circ f(n), P_2^2 \circ f(n))$$

of computable real numbers with the open interval

$$\{x \in \mathbf{R} : P_1^2 \circ f(n) < x < P_2^2 \circ f(n)\}$$

in $\mathbf{R}$.

(4.25) Theorem. *For each $\varepsilon > 0$ there exists an effective sequence $(I_n)_{n=0}^\infty$ of bounded open intervals in $\mathbf{R}$ with rational end points, such that (i) $\mathbf{R}_c \subset \bigcup_{n=0}^\infty I_n$, and (ii) $\sum_{n=0}^N |I_n| < \varepsilon$ for each N.*

Proof.   In this proof we consider $\varphi_0, \varphi_1, \ldots$ to be an effective enumeration of the set of computable partial functions from $\mathbf{N}$ to $\mathbf{Q}$. Choose a positive integer k such that $2^{-k+3} < \varepsilon$. For each pair (m, n) of positive integers set

$$J_{m,n} \equiv (\varphi_m(m + k) - 2^{-m-k+1}, \varphi_m(m + k) + 2^{-m-k+1})$$

if $\mathcal{M}_m$ completes a computation in $n+1$ steps on the input $m+k$; otherwise, set $J_{m,n} \equiv \emptyset$. Note that for each m there is at most one n such that $J_{m,n}$ is nonempty. It is a simple exercise to prove that

$$\mathbf{R}_c \subset \bigcup_{m,n=0}^\infty J_{m,n}.$$

By following the arrows through the diagram below and deleting all occurrences of $\emptyset$, we obtain an effective sequence $I_0, I_1, \ldots$ of open intervals in $\mathbf{R}$ with rational end points such that (i) holds.

$$
\begin{array}{ccccccccc}
J_{0,0} & \rightarrow & J_{0,1} & & J_{0,2} & \rightarrow & J_{0,3} & \cdots \\
& \swarrow & & \nearrow & & \swarrow & \\
J_{1,0} & & J_{1,1} & & J_{1,2} & \cdots \\
\downarrow & \nearrow & & \swarrow & \\
J_{2,0} & & J_{2,1} & \cdots \\
& \swarrow & \\
J_{3,0} & \cdots \\
\downarrow \\
\vdots
\end{array}
$$

On the other hand, for each positive integer N we have

$$\sum_{n=0}^N |I_n| \le \sum_{m,n=0}^\infty |J_{m,n}| \le \sum_{m=0}^\infty 2^{-m-k+2} = 2^{-k+3} < \varepsilon,$$

which proves (ii).   □

At first sight, there is nothing surprising about Theorem (4.25). For, as we have already remarked, $\mathbf{R}_c$ is countable and so can be covered by a sequence of open intervals with rational endpoints and with arbitrarily small total length. However, $\mathbf{R}_c$ is not *effectively* enumerable, so there remains the possibility that we cannot find an *effective* sequence of open intervals that covers $\mathbf{R}_c$ and has arbitrarily small total length. Theorem (4.25) shows that this possibility is not realised.

The following corollary clearly demonstrates the failure of the Heine-Borel theorem in a recursive context.

(4.26) Corollary. *In the notation of* Theorem (4.25), *if* $0 < \varepsilon < 1$, *then the set*

$$C_N \equiv \{x \in \mathbf{R}_c \cap [0,1] : x \notin \bigcup_{n=0}^{N} I_n\}$$

is nonempty for each N.

Proof. It is a simple exercise to express the union of the finitely many intervals $I_0,...,I_N$ as the union of at most N pairwise disjoint open intervals $J_0, \ldots, J_\nu$, each with rational, and therefore computable, end points. Since

$$\sum_{n=0}^{\nu} |J_n| \le \sum_{n=0}^{N} |I_n| < \varepsilon < 1,$$

there exists a point $x \in [0,1]$ such that $x \notin \bigcup_{n=0}^{\nu} J_n$. If x is $0, 1$, or an end point of some interval J_k, then it belongs to C_N. Otherwise, there exists $r > 0$ such that the open interval $(x-r, x+r)$ is contained in both $[0,1]$ and the complement of $\bigcup_{n=0}^{\nu} J_n$, in which case any rational point of $(x-r, x+r)$ belongs to C_N.   □

Since $\mathbf{R}_c \cap [0,1]$ is countable, it has Lebesgue measure 0. Does this destroy all prospect of a recursive development of measure theory for subsets of $\mathbf{R}_c$? It does not: there is such a development in which the recursive measure of $\mathbf{R}_c \cap [0,1]$ is 1, as we would certainly want it to be; see Chapter 3 of [8].

We now investigate the relationship between computability and continuity for functions from $\mathbf{R}_c$ to $\mathbf{R}_c$. We say that a partial function $f : \mathbf{R} \to \mathbf{R}$ is **effectively continuous** if for each $x \in \text{domain}(f)$ there exists a total computable function $h : \mathbf{N} \to \mathbf{N}$ such that if $y \in \text{domain}(f)$, $n \in \mathbf{N}$, and $|x - y| \le 2^{-h(n)}$, then $|f(x) - f(y)| \le 2^{-n}$. On the other hand, we say that f is **effectively uniformly continuous** if the function h can be chosen independent of x; that is, if there exists a total computable function $h : \mathbf{N} \to \mathbf{N}$ such that if $x, y \in \text{domain}(f)$, $n \in \mathbf{N}$, and $|x - y| \le 2^{-h(n)}$, then $|f(x) - f(y)| \le 2^{-n}$.

We state, without proof, the fundamental result relating continuity and computability for functions on $\mathbf{R}_c$—the **Kreisel-Lacombe-Schoenfield-Čeitin Theorem**:

(4.27) Theorem.   *Every total computable function* $f : \mathbf{R}_c \to \mathbf{R}_c$ *is effectively continuous.*

Proof.   For a proof see [8], [11], or [22].   □

Here is an interesting partial converse of this theorem.

(4.28) Proposition.   *Let* $f : \mathbf{R} \to \mathbf{R}$ *be a total function that maps* $\mathbf{R}_c$ *into* $\mathbf{R}_c$ *and is effectively uniformly continuous on* $\mathbf{R}$. *Then the restriction of* f *to* $\mathbf{R}_c$ *is computable.*

Proof.   For each $n \in \mathbf{N}$ let $s_n : \mathbf{R} \to \mathbf{R}$ be the total function that takes the value 1 throughout $[-n, n]$, vanishes outside $[-n-1, n+1]$, and is linear in each of the intervals $[-n - 1, n]$, $[n, n + 1]$. It is straightforward to show that each s_n maps $\mathbf{R}_c$ into $\mathbf{R}_c$, that $(s_n)_{n=1}^{\infty}$ (and therefore $(fs_n)_{n=0}^{\infty}$) is a computable sequence of total computable functions from $\mathbf{R}_c$ to $\mathbf{R}_c$, and that there exists a total computable function $h : \mathbf{N}^2 \to \mathbf{N}$ such that if $x, y \in \mathbf{R}$, if $n, k \in \mathbf{N}$, and if $|x - y| \le 2^{-h(n,k)}$, then $|(fs_n)(x) - (fs_n)(y)| \le 2^{-k}$; the details are left to Exercise (4.29.3). By the Weierstrass Approximation Theorem, there exists a double sequence $(p_{n,k})_{n,k=1}^{\infty}$ of polynomial functions with rational coefficients such that

$$\sup\{|(fs_n)(x) - p_{n,k}(x)| : -n - 2 \le x \le n + 2\} \le 2^{-k-1}$$

for all n and k. Moreover, a close inspection of Bernstein's proof of that theorem shows that the polynomials $p_{n,k}$ can be chosen so that the total function $(n, q, k) \mapsto p_{n,k}(q)$ is computable on $\mathbf{N} \times \mathbf{Q} \times \mathbf{N}$; see Exercise (4.29.5). Now define a computable partial function $\Psi : \mathbf{N}^3 \to \mathbf{Q}$ by

$$
\begin{aligned}
\Psi(n, i, k) \quad &= \quad p_{n,k} \circ \varphi_i \circ h(n, k + 1) \quad && \text{if } h(n, k + 1) \in \text{ domain}(\varphi_i) \text{ and} \\
& && |\varphi_i \circ h(n, k + 1)| \le n + 2, \\
&= \quad 0 \quad && \text{if } h(n, k + 1) \in \text{ domain}(\varphi_i) \text{ and} \\
& && |\varphi_i \circ h(n, k + 1)| > n + 2, \\
&= \quad \text{undefined} \quad && \text{otherwise.}
\end{aligned}
$$

Construct a total computable function $g : \mathbf{N}^2 \to \mathbf{N}$ such that $\varphi_{g(n,i)} = \Psi(n, i, \cdot)$ for all n, i. We prove that if φ_i is a computable real number generator converging to $x \in \mathbf{R}_c$, then $\varphi_{g(n,i)}$ is a computable real number generator converging to $(fs_n)(x)$. To this end, consider any $n, k \in \mathbf{N}$, and first note that

$$|x - \varphi_i \circ h(n, k + 1)| \le 2^{-h(n,k+1)} \le 1;$$

so if $|\varphi_i \circ h(n, k+1)| > n+2$, then $|x| > n+1$ and

$$\varphi_{g(n,i)}(k) = 0 = (fs_n)(x).$$

We may therefore assume that $|\varphi_i \circ h(n, k+1)| \leq n+2$; whence

$$
\begin{aligned}
&\left|\varphi_{g(n,i)}(k) - (fs_n)(x)\right| \\
=\ & |p_{n,k} \circ \varphi_i \circ h(n, k+1) - (fs_n)(x)| \\
\leq\ & |p_{n,k} \circ \varphi_i \circ h(n, k+1) - (fs_n)(\varphi_i \circ h(n, k+1))| \\
& + |(fs_n)(\varphi_i \circ h(n, k+1)) - (fs_n)(x)| \\
\leq\ & 2^{-k-1} + 2^{-k-1} \\
=\ & 2^{-k}.
\end{aligned}
$$

Thus $\varphi_{g(n,i)}$ is a computable real number generator converging to $(fs_n)(x)$. Now define computable partial functions $\alpha, \theta : \mathbf{N} \to \mathbf{N}$ as follows:

$$
\begin{aligned}
\alpha(i) &\equiv \min n \left[|\varphi_i(0)| < n-1 \right], \\
\theta(i) &\equiv g(\alpha(i), i).
\end{aligned}
$$

Consider a computable real number generator φ_i converging to $x \in \mathbf{R}_c$. Since

$$|x| \leq |\varphi_i(0)| + 1 < \alpha(i),$$

we see that

$$\left| f(x) - \varphi_{\theta(i)}(n) \right| = \left| (fs_{\alpha(i)})(x) - \varphi_{g(\alpha(i),i)}(n) \right| \leq 2^{-k}$$

for each k; whence $\varphi_{\theta(i)}$ is a computable real number generator converging to $f(x)$.   □

An elementary theorem of classical analysis states that a continuous function from a compact interval to $\mathbf{R}$ attains its infimum; from which it follows immediately that a continuous, everywhere positive function on a compact interval has positive infimum. In sharp contrast, there is a total computable function $f : \mathbf{R}_c \to \mathbf{R}_c$ that is positive and effectively uniformly continuous on $\mathbf{R}_c$, and whose infimum on $\mathbf{R}_c \cap [-1, 1]$ is 0.

To prove this, construct, as in Theorem (4.25), an effective sequence $(I_n)_{n=0}^{\infty}$ of bounded open intervals in $\mathbf{R}$ with rational end points, such that $\mathbf{R}_c \subset \cup_{n=0}^{\infty} I_n$ and $\sum_{n=0}^{N} |I_n| < 1/2$ for each N. For each n let $t_n : \mathbf{R} \to \mathbf{R}$ vanish outside I_n, equal 1 at the mid-point of I_n, and be linear in each half of I_n. Then (t_n) is a sequence of effectively uniformly continuous total functions from $\mathbf{R}$ to $[0, 1]$ such that for each n, t_n maps $\mathbf{R}_c$ into $\mathbf{R}_c$. It is left to you to show, in Exercise (4.29.6) below, that the sum t of the series $\sum_{n=0}^{\infty} 2^{-n} t_n$ is an effectively uniformly continuous total function on $\mathbf{R}$, and that the restriction of t is a total computable function f from $\mathbf{R}_c$ to $\mathbf{R}_c$. For each $x \in \mathbf{R}_c \cap [-1, 1]$ there exists n such that $x \in I_n$; whence

$$f(x) \geq 2^{-n} t_n(x) > 0.$$

On the other hand, by Corollary (4.26), for each $N \in \mathbf{N}$ there exists a computable real number x_N that belongs to $[-1, 1] \setminus \cup_{n=0}^{N} I_n$ and therefore satisfies

$$f(x_N) = \sum_{n=N+1}^{\infty} 2^{-n} t_n(x_N) \leq \sum_{n=N+1}^{\infty} 2^{-n} = 2^{-N}.$$

Thus $\inf f = 0$.

(4.29) Exercises

.1* Let $f : \mathbf{R} \to \mathbf{R}$ be an effectively continuous partial function that maps $\mathbf{Q} \cap \mathrm{domain}(f)$ into $\mathbf{Q}$. Prove that f maps $\mathbf{R}_c \cap \mathrm{domain}(f)$ into $\mathbf{R}_c$.

.2 Let (f_n) be a sequence of total functions from $\mathbf{R}$ to $\mathbf{R}$ such that for each n, f_n maps $\mathbf{Q}$ into $\mathbf{Q}$, and let g_n be the restriction of f_n to $\mathbf{R}_c$. Suppose that there exists a total computable function $h : \mathbf{N}^2 \to \mathbf{N}$ such that for all n and k, if $|x - y| \leq 2^{-h(n,k)}$, then $|f_n(x) - f_n(y)| \leq 2^{-k}$. Prove that (g_n) is a computable sequence of computable functions from $\mathbf{R}_c$ into $\mathbf{R}_c$.

.3 Under the hypotheses of Proposition (4.28), and using the notation of the proof of that result, prove that each s_n maps $\mathbf{R}_c$ into $\mathbf{R}_c$; that $(s_n)_{n=1}^{\infty}$ is a computable sequence of total computable functions from $\mathbf{R}_c$ to $\mathbf{R}_c$; and that there exists a total computable function $h : \mathbf{N}^2 \to \mathbf{N}$ such that if $x, y \in \mathbf{R}$, if $n, k \in \mathbf{N}$, and if $|x-y| \leq 2^{-h(n,k)}$, then $|(fs_n)(x) - (fs_n)(y)| \leq 2^{-k}$.

.4 Let $(f_n)_{n=1}^{\infty}$ be a sequence of total functions from $\mathbf{R}$ to $\mathbf{R}$ such that each f_n maps $\mathbf{R}_c$ into $\mathbf{R}_c$ and such that (f_n) is a computable sequence of total computable functions from $\mathbf{R}_c$ to $\mathbf{R}_c$. Suppose also that there exists a total computable function $h : \mathbf{N}^2 \to \mathbf{N}$ such that if $x, y \in \mathbf{R}$, $n, k \in \mathbf{N}$, and $|x - y| \leq 2^{-h(n,k)}$, then $|f_n(x) - f_n(y)| \leq 2^{-k}$. Prove that there exists a total computable function $b : \mathbf{N}^2 \to \mathbf{N}$ such that

$$|f_n(x)| \leq b(m, n) \quad (m, n \in N, \ -2^{m-1} \leq x \leq 2^{m-1}).$$

(*Hint*: Consider the values of f_n on a certain finite subset of the interval $[-2^{m-1}, 2^{m-1}]$.)

.5* The following version of the **Weierstrass Approximation Theorem** is proved on pages 18-20 of [31]: *Let f be a continuous mapping* *of* $[0,1]$ *into* $\mathbf{R}$; *let* ε, δ *be positive numbers such that if* $0 \leq x, y \leq 1$ *and* $|x - y| \leq \delta$, *then* $|f(x) - f(y)| \leq \varepsilon$; *and let* $M > 0$ *be such that* $|f(x)| \leq M$ *for all* $x \in [0,1]$. *Then*

$$\left| f(x) - \sum_{i=0}^{n} \binom{n}{i} f(i/n) x^i (1-x)^{n-i} \right| \leq \varepsilon$$

for all $x \in [0,1]$ *and all* $n \geq M/\varepsilon\delta^2$.

Using this information, prove that in the proof of Proposition (4.28) the polynomials $p_{n,k}$ can be chosen such that their coefficients are computable, and such that the total function $(n, q, k) \mapsto p_{n,k}(q)$ is computable on $\mathbf{N} \times \mathbf{Q} \times \mathbf{N}$. (*Hint*: First define total mappings $G : \mathbf{N} \times \mathbf{R} \to \mathbf{R}$ and $H : \mathbf{N} \times \mathbf{R} \to \mathbf{R}$ by

$$G(N, x) \equiv \frac{N+2}{2x-1},$$

$$H(N, x) \equiv \frac{-x}{2(N+2)} - \frac{1}{2}.$$

For all $N, n \in \mathbf{N}$ construct[2] rational numbers $r_{n,i}$ $(0 \leq i \leq n)$ such that the functions $f_{N,n} : \mathbf{R} \to \mathbf{R}$ defined by

$$f_{N,n}(x) \equiv \sum_{i=0}^{n} \binom{n}{i} r_{n,i} x^i (1-x)^{n-i}$$

map $\mathbf{R}_c$ into $\mathbf{R}_c$, $(f_{N,n})_{N,n=0}^{\infty}$ is a computable sequence of total computable functions from $\mathbf{R}_c$ into $\mathbf{R}_c$, and

$$\left| f_{N,n}(x) - \sum_{i=0}^{n} \binom{n}{i} (f s_N) \left(G \left(N, \frac{i}{n} \right) \right) x^i (1-x)^{n-i} \right| \leq 2^{-n}$$

for each $x \in [0,1]$. Next prove that

$$(N, q, n) \mapsto f_{N,n}(q)$$

is a total computable function from $\mathbf{N} \times \mathbf{Q} \times \mathbf{N}$ to $\mathbf{R}_c$. Using Exercise (4.29.4), construct a total computable function $M : \mathbf{N} \to \mathbf{N}$ such that $|(f s_N)(x)| \leq M(N)$ whenever $N \in \mathbf{N}$ and $-N - 2 \leq x \leq N + 2$. Then, using the version of the Weierstrass Approximation Theorem stated above, construct a computable subsequence $(f_{N,n_k})_{N,k=0}^{\infty}$ of $(f_{N,n})$ such that

$$|(f s_N) \circ G(N, x) - f_{N,n_k}(x)| \leq 2^{-k-1} \quad (N, k \in \mathbf{N}, \ x \in [0,1]).$$

---

[2]For this construction, note Exercise (4.24.5).

Finally, set

$$p_{N,k} \equiv f_{N,n_k} \circ H(N, \cdot)$$

for all n, k.)

.6   Let $(t_n)_{n=0}^{\infty}$ be a sequence of effectively uniformly continuous total functions from $\mathbf{R}$ into $[-1, 1]$. Prove that the series $\sum_{n=0}^{\infty} 2^{-n} t_n$ converges effectively and uniformly on $\mathbf{R}$, and that its sum t is an effectively uniformly continuous total function from $\mathbf{R}$ to $\mathbf{R}$. Suppose also that each t_n maps $\mathbf{R}_c$ to $\mathbf{R}_c$; that (t_n) is a computable sequence of total computable functions from $\mathbf{R}_c$ to $\mathbf{R}_c$; and that there exists a total computable function $h : \mathbf{N}^2 \to \mathbf{N}$ such that if $x, y \in \mathbf{R}_c$, if $n, N \in \mathbf{N}$, and if $|x - y| \le 2^{-h(N,n)}$, then $|t_N(x) - t_N(y)| \le 2^{-n}$. Prove that the restriction of t to $\mathbf{R}_c$ is a total computable function from $\mathbf{R}_c$ to $\mathbf{R}_c$.

.7*   With reference to Exercise (4.29.6), complete the details of the example preceding this set of exercises.

.8   Give an example of a total computable function $g : \mathbf{R}_c \to \mathbf{R}_c$ that is effectively continuous, but whose restriction to $\mathbf{R}_c \cap [0, 1]$ is not uniformly continuous. Thus the recursive analogue of the classical uniform continuity theorem is false (cf. (3.16.5) of [15]).

For further information about recursive analysis, see [1], [4], and [26]. A rather different approach to algorithmic aspects of analysis, using a non-classical logic, is found in [5] and Chapter 3 of [8].

5

Rice's Theorem and the Recursion Theorem

In this chapter we turn back from our study of computable real numbers and take a path that will lead to two of the major theorems in computability theory. The first of these, Rice's Theorem, characterises a large class of nonrecursive subsets of $\mathbf{N}$; the second, the Recursion Theorem, has many applications, some of which appear at the end of this chapter, and some, in perhaps unexpected contexts, in later chapters.

Since there is an enumeration of the set of all computable partial functions from $\mathbf{N}$ to $\mathbf{N}$, there is an enumeration of the set of all total computable functions from $\mathbf{N}$ to $\mathbf{N}$. However, as we now prove using another diagonal argument, no enumeration of the set of total computable functions from $\mathbf{N}$ to $\mathbf{N}$ can be effective.

(5.1) Proposition. *If $f_0, f_1, f_2, \ldots$ is an effective enumeration of a sequence of total computable functions from $\mathbf{N}$ to $\mathbf{N}$, then there exists a total computable function $f : \mathbf{N} \to \mathbf{N}$ such that $f \neq f_n$ for each n.*

Proof. Define a total computable function $f : \mathbf{N} \to \mathbf{N}$ by

$$f(n) \equiv f_n(n) + 1.$$

For each n we then have $f(n) \neq f_n(n)$ and therefore $f \neq f_n$. □

(5.2) Corollary. *The set $\{n \in \mathbf{N} : \varphi_n \text{ is total}\}$ is not recursively enumerable and hence is not recursive.* □

(5.3) Exercise

Prove Corollary (5.2).

We now discuss the second of the three questions posed in Chapter 4 at the start of our discussion of computability theory. To this end, we prove that there is no algorithm for deciding whether a given computable partial function on $\mathbf{N}$ is equal to the **identity function id** $: \mathbf{N} \to \mathbf{N}$. This requires a lemma.

(5.4) Lemma.   *If* $\varphi : \mathbf{N} \to \mathbf{N}$ *is a computable partial function, then the partial function* $\Psi : \mathbf{N}^2 \to \mathbf{N}$ *defined by*

$$\begin{aligned}
\Psi(i, j) &= \varphi(j) && \text{if } j \in \text{domain}(\varphi_i) \cap \text{domain}(\varphi), \\
&= \text{undefined} && \text{otherwise}
\end{aligned}$$

is computable.

Proof.   Choose ν so that $\mathcal{M}_\nu$ computes φ. For each $i \in \mathbf{N}$ modify $\mathcal{M}_i$ to create a binary Turing machine $\mathcal{T}_i$ with the following properties. Given an input $j \in \mathbf{N}$, $\mathcal{T}_i$ first replaces j on the left of the tape by $\mathbf{B}j\mathbf{B}j$. $\mathcal{T}_i$ then calls a module that, without affecting the cells to the left of the rightmost instance of j, mimics the action of $\mathcal{M}_i$ on that instance of j. If $j \in \text{domain}(\varphi_i)$, this module will arrive at a configuration in which the tape contains $\mathbf{B}j\mathbf{B}\varphi_i(j)$ and the read/write head is against the leftmost symbol of $\varphi_i(j)$. $\mathcal{T}_i$ then

> deletes $\varphi_i(j)$;
> moves each unit of j one place to the left on the tape, leaving
>   blanks everywhere else;
> places the read/write head against the leftmost cell; and
> calls a module that mimics the action of $\mathcal{M}_\nu$ on j.

It is easy to see that $\mathcal{T}_i$ computes the partial function Ψ.   □

(5.5) Theorem.   *The set* $\{i \in \mathbf{N} : \varphi_i = \mathbf{id}\}$ *is not recursive.*

Proof.   By Lemma (5.4) and the *s-m-n* theorem, there exists a total computable function $s : \mathbf{N} \to \mathbf{N}$ such that

$$\begin{aligned}
\varphi_{s(i)}(j) &= j && \text{if } j \in \text{domain}(\varphi_i), \\
&= \text{undefined} && \text{otherwise.}
\end{aligned}$$

Suppose the total function $f : \mathbf{N} \to \mathbf{N}$, defined by

$$\begin{aligned}
f(i) &= 1 && \text{if } \varphi_i = \mathbf{id}, \\
&= 0 && \text{otherwise,}
\end{aligned}$$

is computable. Then the composite function $f \circ s : \mathbf{N} \to \{0, 1\}$ is total and computable. But

$$\begin{aligned}
f(s(i)) = 1 \quad &\Leftrightarrow \quad \varphi_{s(i)} = \mathbf{id}, \\
&\Leftrightarrow \quad \varphi_i \text{ is total.}
\end{aligned}$$

It follows that $f \circ s$ is the characteristic function of

$$\{i \in \mathbf{N} : \varphi_i \text{ is total}\},$$

which is therefore a recursive set. This contradicts Corollary (5.2).   □

We can now prove the unsolvability of the equivalence problem.

(5.6) Corollary.   *There is no total computable function* $F : \mathbf{N}^2 \to \{0, 1\}$ *such that* $F(i, j) = 1$ *if and only if* $\varphi_i = \varphi_j$.

Proof.   Suppose such a function F exists. Then, choosing j such that $\mathbf{id} = \varphi_j$, we see that the total computable function $i \mapsto F(i, j)$ on $\mathbf{N}$ is the characteristic function of $\{i \in \mathbf{N} : \varphi_i = \mathbf{id}\}$. This contradicts Theorem (5.5).   □

(5.7) Exercises

.1   Define the total function **stat**: $\mathbf{N} \to \mathbf{N}$ by

$$\mathbf{stat}(n) \equiv \text{ the number of states in } \mathcal{M}_n.$$

Why is **stat** computable? Prove that the total function $f : \mathbf{N} \to \mathbf{N}$ defined by

$$f(n) \equiv \min\{\mathbf{stat}(k) : \varphi_k = \varphi_n\}$$

is not computable. (*Hint*: What partial functions are computed by normalised binary Turing machines with exactly one state?)

.2* Prove that the total function **lindex** : $\mathbf{N} \to \mathbf{N}$ defined by

$$\mathbf{lindex}(n) \equiv \min\{i \in \mathbf{N} : \varphi_i = \varphi_n\}$$

is not computable.[1]

.3   Define a total function $s : \mathbf{N}^2 \to \mathbf{N}$ by

$$s(k, n) \equiv \min\{i : \varphi_i = \varphi_k^{(2)}(\cdot, n)\}.$$

Prove that s is not computable (cf. the remarks preceding the statement of the *s-m-n* theorem in Chapter 3).

---

[1] A similar function is used by Chaitin in his work on LISP program-size complexity [12].

At first sight, our next lemma may be rather surprising.

(5.8) Lemma.   *Let $\epsilon : \mathbf{N} \to \mathbf{N}$ be the empty partial function, and $\varphi : \mathbf{N} \to \mathbf{N}$ a computable partial function with nonempty domain. Then there is a total computable function $f : \mathbf{N} \to \mathbf{N}$ such that*

$$\varphi_{f(i)} \;=\; \varphi \quad \text{if } i \in K,$$
$$\;=\; \epsilon \quad \text{otherwise.}$$

Proof.   Consider any $i \in \mathbf{N}$, and design a Turing machine $\mathcal{T}_i$ as follows. Given the (unary form of) n as input, $\mathcal{T}_i$ first moves each unit of n one place to the right, leaving $\mathbf{B}$ in the leftmost cell. It then

writes $\mathbf{B}i$ on the right of the rightmost unit of n;
places the read/write head against the leftmost symbol of i;
without affecting the cells to the left of i, simulates the action of $\mathcal{M}_i$
  on the input i, but replaces the halt state of $\mathcal{M}_i$ by a Turing machine
  module that
    moves each unit of n one cell to the left, leaving blanks
      everywhere else on the tape,
    moves the read/write head to the leftmost cell, and
    simulates a Turing machine that computes φ.

It is clear that we can construct $\mathcal{T}_i$ so that it is normalised and depends uniquely on i. Thus the total, and clearly computable, mapping f that carries i to the index of $\mathcal{T}_i$ has the desired properties. $\square$

(5.9) Exercise

*   Let φ be a computable partial function with nonempty domain, and define a computable partial function $\Psi : \mathbf{N}^2 \to \mathbf{N}$ such that

$$\Psi(i,j) \;=\; \varphi(j) \qquad \text{if } i \in K,$$
$$\;=\; \text{undefined} \quad \text{otherwise.}$$

Use this function to give an alternative proof of Lemma (5.8).

The arguments we have used to prove many of the results of this and the previous section provide a good grounding in the techniques of computability theory. However, several of those results, and many others, can be obtained as simple consequences of the following very general theorem.

(5.10) Rice's Theorem.   *If I is a nonempty proper[2] recursive subset of $\mathbf{N}$, then there exist i, j such that $i \in I$, $j \in \mathbf{N} \backslash I$, and $\varphi_i = \varphi_j$.*

---

[2]Recall that a subset S of a set X is **proper** if $S \neq \emptyset$ and $S \neq X$.

Proof.   Suppose the contrary, so that $\varphi_i \neq \varphi_j$ whenever $i \in I$ and $j \in \mathbf{N}\backslash I$. Interchanging I and $\mathbf{N}\backslash I$ if necessary, we may assume that I contains an index (and therefore all indices) of some computable partial function φ with nonempty domain, and that $\mathbf{N}\backslash I$ contains an index (and therefore all indices) of the empty function ϵ. Define the total function $f : \mathbf{N} \to \mathbf{N}$ as in Lemma (5.8), and let χ be the characteristic function of I. Then $\chi \circ f : \mathbf{N} \to \{0,1\}$ is a total computable function. If $\varphi_{f(i)} = \epsilon$, then $f(i)$, being an index of ϵ, belongs to $\mathbf{N}\backslash I$; so if $f(i) \in I$, then $\varphi_{f(i)} \neq \epsilon$ and therefore $\varphi_{f(i)} = \varphi$. On the other hand, if $\varphi_{f(i)} = \varphi$, then $f(i)$, being an index of φ, belongs to I. Hence

$$\chi(f(i)) = 1 \quad \Leftrightarrow \quad \varphi_{f(i)} = \varphi$$
$$\Leftrightarrow \quad i \in K.$$

Since $\chi \circ f$ is computable, K is recursive. This contradicts Corollary (4.3).
□

Bearing in mind the effective indentification of a normalised binary Turing machine with its index, we see that Rice's Theorem admits the following interpretation: if P is a decidable property that holds for some but not all normalised binary Turing machines, then there are Turing machines $\mathcal{M}_i$ and $\mathcal{M}_j$, one with the property P and one without P, that compute the same partial function; so the information we need in order to decide whether or not a normalised binary Turing machine $\mathcal{M}$ has the property P is not provided solely by the partial function computed by $\mathcal{M}$.

We say that a subset I of $\mathbf{N}$ **respects indices** if $j \in I$ whenever $\varphi_j = \varphi_i$ for some $i \in I$ (or, equivalently, if $\varphi_i \neq \varphi_j$ whenever $i \in I$ and $j \notin I$). This definition leads immediately to a useful re-expression of Rice's Theorem.

(5.11) Corollary.   *A nonempty proper subset of* $\mathbf{N}$ *that respects indices is not recursive.*   □

This corollary enables us to prove the undecidability of the problems associated in the obvious way with the following subsets of $\mathbf{N}$, each of which respects indices:

$\{i : \varphi_i = \varphi_j\}$, where j is a given natural number,

$\{i : \varphi_i \text{ is total}\}$,

$\{i : a \in \text{domain}(\varphi_i)\}$, where a is a given natural number,

$\{i : \varphi_i \text{ is a constant function}\}$,

$\{i : \text{domain}(\varphi_i) \text{ is finite}\}$.

The undecidability of the first two of these problems also follows from Corollaries (5.6) and (5.2), respectively.

Note that the undecidability of the halting problem is not a direct consequence of Rice's Theorem: for, as the next theorem will enable us to prove, K does not respect indices.

(5.12) Exercise

Without using Rice's Theorem, prove that the decision problems associated with the sets

(i) $\{i : a \in \operatorname{domain}(\varphi_i)\}$, where a is a given natural number, and
(ii) $\{i : \varphi_i$ is a constant function$\}$,

are undecidable. (*Hint for* (i): Consider the partial function $\Psi : \mathbf{N}^2 \to \mathbf{N}$, where $\Psi(m, n) = 1$ if $n = a$ and $m \in K$, and $\Psi(m, n)$ is undefined otherwise.)

Each computable partial function $\theta : \mathbf{N} \to \mathbf{N}$ gives rise to an associated sequence of computable partial functions from $\mathbf{N}^n$ to $\mathbf{N}$: namely,

$$\varphi_{\theta(0)}^{(n)}, \ \varphi_{\theta(1)}^{(n)}, \ \varphi_{\theta(2)}^{(n)}, \dots$$

(Recall that $\varphi_{\theta(k)}^{(n)} \equiv \mathcal{U}_n(\theta(k), \cdot)$, where $\mathcal{U}_n$ is the universal Turing machine for n-ary computable partial functions, discussed on page 42; so if $\theta(k)$ is undefined, then $\varphi_{\theta(k)}^{(n)}$ is the empty partial function from $\mathbf{N}^n$ to $\mathbf{N}$.) The s-m-n theorem shows that the mapping $k \mapsto \varphi_{\theta(k)}^{(n)}$ is an effective enumeration of its range. It follows that the effective enumerations of sets of computable partial functions from $\mathbf{N}^n$ to $\mathbf{N}$ are precisely the listings of the form

$$\varphi_{\varphi_k(0)}^{(n)}, \ \varphi_{\varphi_k(1)}^{(n)}, \ \varphi_{\varphi_k(2)}^{(n)}, \dots \tag{5.1}$$

where $k \in \mathbf{N}$.

Note that for each k, the kth term of the **diagonal effective enumeration**

$$\varphi_{\varphi_0(0)}^{(n)}, \ \varphi_{\varphi_1(1)}^{(n)}, \ \varphi_{\varphi_2(2)}^{(n)}, \dots \tag{5.2}$$

is the same as the kth term in the sequence (5.1).

(5.13) The Recursion Theorem.   *For each $n \in \mathbf{N}$, and each total computable function $f : \mathbf{N} \to \mathbf{N}$, there exists $i \in \mathbf{N}$ such that $\varphi_i^{(n)} = \varphi_{f(i)}^{(n)}$.*

Proof.   To each total computable function $f : \mathbf{N} \to \mathbf{N}$ there corresponds a natural effective enumeration derived from the diagonal one (5.2)—namely,

$$\varphi_{f \circ \varphi_0(0)}^{(n)}, \ \varphi_{f \circ \varphi_1(1)}^{(n)}, \ \varphi_{f \circ \varphi_2(2)}^{(n)}, \dots$$

Choosing an index ν of the mapping $k \mapsto f \circ \varphi_k(k)$, we see from the remark immediately preceding this theorem that $\varphi_{\varphi_\nu(\nu)}^{(n)} = \varphi_{f \circ \varphi_\nu(\nu)}^{(n)}$; so it appears that we can complete the proof by taking $i \equiv \varphi_\nu(\nu)$. However, this will not quite do, as we have no guarantee that $\varphi_\nu(\nu)$ is defined. To get round this obstacle, we invoke the s-m-n theorem, to obtain a *total* computable function $s : \mathbf{N} \to \mathbf{N}$ such that $\varphi_{s(i)}^{(n)} = \varphi_{f \circ \varphi_i(i)}^{(n)}$ for each i. We then choose an index m of s and set $i \equiv \varphi_m(m)$ to obtain

$$\varphi_i^{(n)} = \varphi_{\varphi_m(m)}^{(n)} = \varphi_{s(m)}^{(n)} = \varphi_{f \circ \varphi_m(m)}^{(n)} = \varphi_{f(i)}^{(n)}. \qquad \square$$

Taking $n = 1$ and $f(k) = k + 1$ in Theorem (5.13), we immediately see that there exists i such that $\varphi_i = \varphi_{i+1}$. The proof of the following partial generalisation of this result provides a more illuminating example of the application of the Recursion Theorem: *for each computable partial function $\varphi : \mathbf{N} \to \mathbf{N}$ and each positive integer k, there exists i such that if $\varphi_i(n)$ is defined, then so are $\varphi(n)$ and $\varphi_{i+j}(n)$ $(1 \le j \le k)$, and*

$$\varphi_i(n) = \varphi_{i+1}(n) = \cdots = \varphi_{i+k}(n) = \varphi(n).$$

To prove this, given φ and k, define a computable partial function $\Psi : \mathbf{N}^2 \to \mathbf{N}$ by

$$\Psi(m, n) \equiv \varphi_m(n) + \sum_{j=m}^{m+k} |\varphi_j(n) - \varphi(n)|.$$

By the s-m-n theorem, there exists a total computable function $s : \mathbf{N} \to \mathbf{N}$ such that $\Psi(m, \cdot) = \varphi_{s(m)}$ for all m. Applying the Recursion Theorem, compute i such that $\varphi_i = \varphi_{s(i)}$; then

$$\varphi_i = \varphi_i + \sum_{j=i}^{i+k} |\varphi_j - \varphi|.$$

If $\varphi_i(n)$ is defined, then so are $\varphi(n)$ and $\varphi_{i+j}(n)$ $(1 \le j \le k)$,

$$\sum_{j=i}^{i+k} |\varphi_j(n) - \varphi(n)| = 0,$$

and therefore

$$\varphi_i(n) = \varphi_{i+1}(n) = \cdots = \varphi_{i+k}(n) = \varphi(n).$$

The procedure used in the above illustration is typical of many applications of the Recursion Theorem: we first define an appropriate computable partial function Ψ of two variables, then use the s-m-n theorem to "pull back" to functions $\varphi_{s(m)}$ of one variable, and finally apply the Recursion Theorem to the total computable function s. This procedure is used in several of the next set of exercises.

(5.14) Exercises

.1* For a given natural number n draw the state diagram of a Turing machine $\mathcal{T}_n$ that computes the partial function $\varphi : \mathbf{N} \to \mathbf{N}$ defined by

$$\varphi(i) \quad = \quad 1 \qquad\qquad \text{if } i = n,$$
$$= \quad \text{undefined} \quad \text{otherwise.}$$

Your solution to this exercise should provide an algorithm for constructing $\mathcal{T}_n$ uniquely from n.

.2 Prove that there exists an index n such that $\text{domain}(\varphi_n) = \{n\}$. (*Hint:* Use either Exercise (5.14.1) or the s-m-n theorem; then apply the Recursion Theorem.) Use this to prove that K does not respect indices.

.3 Prove that (i) there exists an index i such that $i \in \text{domain}(\varphi_i) = K$, and (ii) there exists an index j such that $j \notin \text{domain}(\varphi_j) = K$.

.4 Given a proper recursive subset I of $\mathbf{N}$, choose i in I and j in $\mathbf{N} \backslash I$, and define a total function $f : \mathbf{N} \to \mathbf{N}$ by

$$f(n) \quad = \quad j \quad \text{if } n \in I,$$
$$= \quad i \quad \text{if } n \notin I.$$

Show that f is computable. Applying the Recursion Theorem to f, give another proof of Rice's Theorem. Can you see any advantage of this proof over the previous one?

.5 Use the Recursion Theorem to prove that K is not recursive.

.6* Suppose there is a total computable function $f : \mathbf{N} \to \{0, 1\}$ such that for each m,

$$f(m) = 1 \Leftrightarrow \text{domain}(\varphi_m) \neq \emptyset.$$

Define a partial function $\Psi : \mathbf{N}^2 \to \mathbf{N}$ as follows: for all $m, n \in \mathbf{N}$,

$$\Psi(m, n) \quad = \quad 1 \qquad\qquad \text{if } f(m) = 0,$$
$$= \quad \text{undefined} \quad \text{if } f(m) = 1.$$

Use the s-m-n theorem and the Recursion Theorem to deduce a contradiction.

.7 Let $f : \mathbf{N} \dashrightarrow \mathbf{N}$ be a total computable function. Prove that there are infinitely many values of i such that $\varphi_i = \varphi_{f(i)}$.

.8 In the Recursion Theorem we cannot choose the index i such that φ_i is *total* and $\varphi_i = \varphi_{f(i)}$: to see this, find an example of a total computable function $f : \mathbf{N} \to \mathbf{N}$ such that if φ_n is total, then $\varphi_{f(n)}$ is total and distinct from φ_n.

.9 Prove the **Extended Recursion Theorem**: *For each $n \geq 1$ there exists a total computable function $t : \mathbf{N} \to \mathbf{N}$ such that if $\varphi_k^{(n)}$ is total, then*

$$\varphi_{t(k)}^{(n)} = \varphi_{\varphi_k(t(k))}^{(n)}.$$

(Thus, in the notation of Theorem (5.13), the index i can be obtained as a computable function of an index of f.)

.10 Recalling the definition of an *acceptable programming system* (page 45), prove the following generalisation of the Recursion Theorem: *For each acceptable programming system $\psi_0, \psi_1, \dots$ and each total computable function $f : \mathbf{N} \to \mathbf{N}$ there exists $i \in \mathbf{N}$ such that $\psi_i = \psi_{f(i)}$.*

.11 Give examples of the following:

(a) An acceptable programming system $\psi_0, \psi_1, \dots$ such that there does not exist i with $\psi_i = \psi_{i+1} = \psi_{i+2}$.

(b) An acceptable programming system $\psi_0, \psi_1, \dots$ such that for each computable partial function $\varphi : \mathbf{N} \to \mathbf{N}$ and each positive integer k there exists i such that $\varphi_i = \varphi_{i+1} = \cdots = \varphi_{i+k} = \varphi$.

What do these examples, taken with Exercise (5.14.10), tell you about the application of the Recursion Theorem given before this set of exercises?

.12 Prove the **Double Recursion Theorem**: *If $F, G : \mathbf{N}^2 \to \mathbf{N}$ are total computable functions, then there exist i, j such that $\varphi_i = \varphi_{F(i,j)}$ and $\varphi_j = \varphi_{G(i,j)}$.* (*Hint:* First show that there exists a total computable function $h : \mathbf{N} \to \mathbf{N}$ such that $\varphi_{F(h(i),i)} = \varphi_{h(i)}$ for each i.)

Since our proof of the Recursion Theorem depends on neither (i) the nonrecursiveness of K nor (ii) Rice's Theorem, and since, according to Exercises (5.14.4,5), both (i) and (ii) can be derived as consequences of the Recursion Theorem, we see that the name of that theorem properly reflects its status as perhaps the central result of computability theory.

We round off this discussion of the Recursion Theorem with a mischievous application, the construction of a self-replicating virus—that is, a Turing machine $\mathcal{M}$ that, when given any natural number as input, completes a computation that outputs the same Turing machine $\mathcal{M}$, in encoded form, on the tape.[3] To obtain a formal definition of this notion, recall the encoding function γ defined before Theorem (3.6), and let $F : \mathbf{N} \to \mathbf{N}$ be the composition of the function $n \mapsto \mathcal{M}_n$ with γ. By a **self-replicating virus** we mean a normalised binary Turing machine $\mathcal{M}_\nu$ such that $\varphi_\nu(n) = F(\nu)$ for each $n \in \mathbf{N}$.

Since F is computable, so is the partial function $(i, j) \mapsto F \circ \varphi_i(j)$ on $\mathbf{N}^2$. By the *s-m-n* theorem, there exists a total computable function $s : \mathbf{N} \to \mathbf{N}$ such that $\varphi_{s(i)} = F \circ \varphi_i$ for each i. Another application of the *s-m-n* theorem yields a total computable function $f : \mathbf{N} \to \mathbf{N}$ such that $\varphi_{f(i)}(n) = s(i)$ for all $i, n \in \mathbf{N}$. By the Recursion Theorem, there exists an index i such that $\varphi_i = \varphi_{f(i)}$; set $\nu \equiv s(i)$ for this i. Then for all n we have

$$\varphi_\nu(n) = \varphi_{s(i)}(n) = F \circ \varphi_i(n) = F \circ \varphi_{f(i)}(n) = F \circ s(i) = F(\nu).$$

Thus $\mathcal{M}_\nu$ is a self-replicating virus.

(5.15) Exercise

*   Prove that

$$\{n \in \mathbf{N} : \mathcal{M}_n \text{ is a self-replicating virus}\}$$

is not a recursive subset of $\mathbf{N}$.

Rice's Theorem characterises the nontrivial recursive subsets of $\mathbf{N}$. In the remainder of this chapter we show how Rice's Theorem can be extended in several ways to provide necessary conditions for a subset of $\mathbf{N}$ to be recursively enumerable.

We begin with a generalisation of Lemma (5.8).

(5.16) Lemma.

If $\varphi, \psi : \mathbf{N} \to \mathbf{N}$ are computable partial functions with $\psi \subset \varphi$, then there exists a total computable function $f : \mathbf{N} \to \mathbf{N}$ such that for all $m \in \mathbf{N}$,

$$\varphi_{f(m)} \quad = \quad \varphi \quad \text{if } m \in K,$$
$$= \quad \psi \quad \text{if } m \notin K.$$

Proof.   Define a total computable function $H : \mathbf{N}^3 \to \mathbf{N}$ by

$$H(i, n, k) \quad = \quad 1 \quad \text{if } \mathcal{M}_i \text{ computes } \varphi_i(n) \text{ in } k + 1 \text{ steps,}$$
$$= \quad 0 \quad \text{otherwise.}$$

---

[3] An informal discussion of two types of computer virus, including the one discussed here, is found in [16].

Choosing an index ν for ψ, next define a computable partial function Ψ : $\mathbf{N}^3 \to \mathbf{N}$ by

$$
\begin{aligned}
\Psi(m, n, k) &= \psi(n) && \text{if } H(\nu, n, k) = 1 \text{ and} \\
& && H(m, n, j) = 0 \text{ for } 0 \le j < k, \\
&= \varphi(n) && \text{if } H(m, m, k) = 1 \text{ and} \\
& && H(\nu, n, j) = 0 \text{ for } 0 \le j < k, \\
&= \text{undefined} && \text{otherwise.}
\end{aligned}
$$

As $\psi \subset \varphi$, this is an unambiguous definition of a function. Noting that, for given m and n, there is at most one value of k such that $(m, n, k) \in \text{domain}(\Psi)$, define computable partial functions $\sigma : \mathbf{N}^2 \to \mathbf{N}$ and $\theta : \mathbf{N}^2 \to \mathbf{N}$ as follows:

$$
\begin{aligned}
\sigma(m, n) &= k && \text{if } (m, n, k) \in \text{domain}(\Psi), \\
&= \text{undefined} && \text{otherwise}
\end{aligned}
$$

and

$$
\theta(m, n) \equiv \Psi(m, n, \sigma(m, n)).
$$

Then

$$
\begin{aligned}
\theta(m, n) &= \varphi(n) && \text{if } m \in K, \\
&= \psi(n) && \text{if } m \notin K.
\end{aligned}
$$

An application of the s-m-n theorem completes the proof. $\square$

(5.17) Exercises

.1   Explain why the definition of the function Ψ in the above proof is unambiguous and why Ψ is computable.

.2   Does Lemma (5.16) hold without the hypothesis that $\psi \subset \varphi$?

Recall from Exercise (4.5.2) that the subset $\bar{K}$ of $\mathbf{N}$ is not recursively enumerable.

(5.18) Proposition.   *Let I be a recursively enumerable subset of $\mathbf{N}$ that respects indices. If $i \in I$, $j \in \mathbf{N}$, and $\varphi_i \subset \varphi_j$, then $j \in I$.*

Proof.   We may assume that I is nonempty. By Proposition (3.2), there exists ν such that $I = \text{domain}(\varphi_\nu)$. Consider $i \in I$ and $j \in \mathbf{N}$ such that $\varphi_i \subset \varphi_j$. According to Lemma (5.16), there exists a total computable function $f : \mathbf{N} \to \mathbf{N}$ such that for all $m \in \mathbf{N}$,

$$
\begin{aligned}
\varphi_{f(m)} &= \varphi_j && \text{if } m \in K, \\
&= \varphi_i && \text{if } m \notin K.
\end{aligned}
$$

Suppose that $j \notin I$. Since I respects indices, we see that if $m \notin K$, then $f(m) \in I$, and that if $m \in K$, then $f(m) \notin I$. It follows that the domain of the computable partial function $\varphi_\nu \circ f : \mathbf{N} \to \mathbf{N}$ is $\bar{K}$, which is therefore recursively enumerable, by Theorem (3.3). This contradiction shows that $j \in I$.   □

Proposition (5.18) deals with extensions of computable partial functions; the next proposition, which will be used in the proof of our extension (5.27) of Rice's Theorem, deals with finite restrictions of computable partial functions.

If φ, ψ are partial functions on $\mathbf{N}$ such that domain(φ) is finite and $\varphi \subset \psi$, then we say that φ is a **finite subfunction** of ψ.

(5.19) Proposition.   *If I is a recursively enumerable subset of $\mathbf{N}$ that respects indices, then for each $n \in I$ there exists $i \in I$ such that φ_i is a finite subfunction of φ_n.*

Proof.   Given $n \in I$, first apply the *s-m-n* theorem to construct a total computable function $s : \mathbf{N} \to \mathbf{N}$ such that for all i and j,

$$
\begin{aligned}
\varphi_{s(i)}(j) &= \text{undefined} \quad \text{if } \mathcal{M}_i \text{ computes } \varphi_i(i) \text{ in at} \\
&\qquad\qquad\qquad \text{most } j+1 \text{ steps,} \\
&= \varphi_n(j) \qquad\;\; \text{otherwise.}
\end{aligned}
$$

Note that if $i \in \bar{K}$, then $\varphi_{s(i)} = \varphi_n$, and therefore, since I respects indices, $s(i) \in I$. Suppose that if φ_k is a finite subfunction of φ_n, then $k \notin I$. If $i \in K$, then $\varphi_{s(i)}$ is a finite subfunction of φ_n, so $s(i) \notin I$. Hence $i \in \bar{K}$ if and only if $s(i) \in I$. Using Proposition (3.2) to construct a computable partial function θ with domain I, we now see that domain$(\theta \circ s) = \bar{K}$. Since $\theta \circ s$ is computable, it follows from Theorem (3.3) that $\bar{K}$ is recursively enumerable, a contradiction.   □

Although the stated form of Proposition (5.19) is sufficient for most applications, there is an interesting strong version of that theorem whose proof we leave as a (hard!) exercise.

(5.20) Proposition.   *There exists a total computable function $f : \mathbf{N}^2 \to \mathbf{N}$ with the following property: if domain(φ_m) respects indices, then for all $n \in$ domain(φ_m), $f(m,n) \in$ domain(φ_m) and $\varphi_{f(m,n)}$ is a finite subfunction of φ_n.*   □

(5.21) Exercises

.1   Prove that $S \equiv \{i \in \mathbf{N} : \varphi_i = \epsilon\}$ is not recursively enumerable. Is $\mathbf{N} \backslash S$ recursively enumerable?

.2* For which computable partial functions $\varphi : \mathbf{N} \to \mathbf{N}$ is $\{n \in \mathbf{N} : \varphi_n = \varphi\}$ recursively enumerable?

.3   Use Proposition (5.18) to give another proof of Rice's Theorem.

.4* Prove that neither $S \equiv \{i \in \mathbf{N} : \varphi_i \text{ is total}\}$ nor $\mathbf{N} \backslash S$ is recursively enumerable. Thus, in a very strong sense, there is no algorithm for deciding whether or not a computable partial function is total.

.5   Prove Proposition (5.20).

We now describe an encoding of finite subsets of $\mathbf{N}$ as natural numbers. Consider any element S of the set $\mathcal{F}$ of all finite subsets of $\mathbf{N}$. If S is empty, define $\mu(S) \equiv 0$; if S is nonempty, let $n_0, \ldots, n_k$ be the elements of S written in a strictly increasing finite sequence, and define $\mu(S)$ to be the binary integer $u_0 0 u_1 0 \ldots 0 u_k$, where for each j, u_j is the unary form of n_j. Then μ is a one-one mapping of $\mathcal{F}$ into $\mathbf{N}$. We shall identify S with $\mu(S)$ when it is convenient to do so.

By a **computable partial function from** N **into** $\mathcal{F}$ we shall mean a partial function $\varphi : \mathbf{N} \to \mathcal{F}$ such that the corresponding partial function $\mu \circ \varphi : \mathbf{N} \to \mathbf{N}$ is computable. In other words, φ is computable if and only if it is computable when considered as a mapping that carries a natural number into the encoding of a finite subset of $\mathbf{N}$.

(5.22) Exercises

.1   Find the subset of $\mathbf{N}$ whose encoded form is 11011110111111.

.2   Prove that the mapping $n \mapsto \{n, n^2\}$ of $\mathbf{N}$ into $\mathcal{F}$ is computable.

For our extensions of Rice's Theorem we require a special effective enumeration of the set of computable partial functions from $\mathbf{N}$ to $\mathbf{N}$ with finite domain.

(5.23) Lemma.   *There exist a one-one effective enumeration*

$$\psi_0, \psi_1, \ldots$$

of the set of all computable partial functions from $\mathbf{N}$ *to* $\mathbf{N}$ *with finite domain, and a total computable function* $d : \mathbf{N} \to \mathcal{F}$, *such that*

(i) $d(n) = \operatorname{domain}(\psi_n)$ *for each* n,

(ii) $\{n \in \mathbf{N} : d(n) \neq \emptyset\}$ *is a recursive set*.

Proof.   Define a computable partial function $\Psi : \mathbf{N}^3 \to \mathbf{N}$ by

$$\Psi(m, n, k) \;=\; \varphi_m(k) \qquad \text{if } k \leq n \text{ and } \mathcal{M}_m \text{ computes } \varphi_m(k)$$
$$\text{in at most } n+1 \text{ steps,}$$
$$\;=\; \text{undefined} \quad \text{otherwise.}$$

By running $\mathcal{M}_m$ for at most $n+1$ steps on each of the inputs $0, \ldots, n$, we can compute, from the pair (m, n), the (code for the) finite domain $D_{m,n}$ of $\Psi(m, n, \cdot)$. Also, we can decide, for all m, m', n, and n', whether the computable partial functions $\Psi(m, n, \cdot)$ and $\Psi(m', n', \cdot)$ from $\mathbf{N}$ to $\mathbf{N}$ are equal. Following the arrows through the diagram below and deleting all repetitions, we obtain a one-one total computable function

$$ n \mapsto (\psi_n, \mathrm{domain}(\psi_n)) $$

on $\mathbf{N}$, where $\psi_n : \mathbf{N} \to \mathbf{N}$ is a computable partial function and $\mathrm{domain}(\psi_n)$ is finite; moreover, each computable partial function from $\mathbf{N}$ to $\mathbf{N}$ with finite domain equals ψ_n for exactly one $n \in \mathbf{N}$.

$$
\begin{array}{llll}
(\Psi(0,0,\cdot), D_{0,0}) & \to & (\Psi(0,1,\cdot), D_{0,1}) & (\Psi(0,2,\cdot), D_{0,2}) & \to & \cdots \\
& \swarrow & & \nearrow & \\
(\Psi(1,0,\cdot), D_{1,0}) & & (\Psi(1,1,\cdot), D_{1,1}) & \cdots \\
\downarrow & \nearrow & \\
(\Psi(2,0,\cdot), D_{2,0}) & \cdots \\
\vdots
\end{array}
$$

To show that the enumeration $\psi_0, \psi_1, \ldots$ is effective, we simply apply the s-m-n theorem to the (informally) computable partial function $(n, k) \mapsto \psi_n(k)$ on $\mathbf{N}^2$. Finally, we describe an algorithm for deciding, for a given n, whether $d(n)$ is empty. First compute i, j such that $\psi_n = \Psi(i, j, \cdot)$. Then check whether $\mathcal{M}_i$ completes a computation in at most $n+1$ steps on any of the inputs $0, \ldots, j$. If it does, then $d(n)$ is nonempty; otherwise, $d(n)$ is empty. $\square$

From now on, we shall take the effective enumeration $\psi_0, \psi_1, \ldots$ of the set of computable partial functions from $\mathbf{N}$ to $\mathbf{N}$ with finite domain, and the mapping $d : \mathbf{N} \to \mathbf{N}$, as in Lemma (5.23). We shall also let $\mathbf{trans}$ be a total computable function from $\mathbf{N}$ to $\mathbf{N}$ such that $\psi_n = \varphi_{\mathbf{trans}(n)}$ for all $n \in \mathbf{N}$; as noted in the proof of Lemma (5.23), the existence of $\mathbf{trans}$ is a consequence of the s-m-n theorem.

(5.24) Exercises

.1   Prove that $J \equiv \{j \in \mathbf{N} : \mathrm{domain}(\psi_j) \neq \emptyset\}$ is recursively enumerable.

.2   Let $\theta : \mathbf{N} \to \mathbf{N}$ be a computable partial function, and define a partial function $\psi : \mathbf{N} \to \mathbf{N}$ by

$$
\begin{aligned}
\psi(n) &= \max d(\theta(n)) \quad \text{if } n \in \mathrm{domain}(\theta) \text{ and } d(\theta(n)) \neq \emptyset, \\
&= \text{undefined} \quad \text{otherwise.}
\end{aligned}
$$

Explain why ψ is computable.

.3   For this exercise we recall the **Goldbach Conjecture**:

> **GC**   *Every even integer ≥ 4 is the sum of two primes.*

(Nobody knows if this conjecture is true.) Consider the following definition of a set S :

$$
\begin{aligned}
S &= \emptyset && \text{if GC is true,} \\
&= \{1\} && \text{if GC is false.}
\end{aligned}
$$

What is wrong with the following argument? Since S is finite, there exists n such that $S = d(n)$; according to Lemma (5.23), we can decide whether or not $d(n)$ is empty; so we can either prove or disprove the Goldbach Conjecture.

It is important to realise that, as the following exercises show, there is no computable partial function θ with the following property: if φ_i has finite domain, then $\theta(i)$ is defined and $\varphi_i = \psi_{\theta(i)}$. In other words, there is no algorithm which, applied to any computable partial function φ with finite domain, enables us to find the unique position of φ in the list $\psi_0, \psi_1, \ldots$

(5.25) Exercises

In these two exercises, $F \equiv \{i \in \mathbf{N} : \text{domain}(\varphi_i) \text{ is finite}\}$.

.1   Let $\theta : \mathbf{N} \to \mathbf{N}$ be a computable partial function whose domain includes F. Prove that there exists $n \in F$ such that $\text{domain}(\varphi_n)$ is both nonempty and disjoint from $\text{domain}(\psi_{\theta(n)})$. Hence prove that there is no computable partial function $\gamma : \mathbf{N} \to \mathbf{N}$ such that

> (i) $F \subset \text{domain}(\gamma)$ and
> (ii) $\varphi_n = \psi_{\gamma(n)}$ for each $n \in F$.

.2   Let $\theta : \mathbf{N} \to \mathbf{N}$ be a computable partial function whose domain includes F. Prove that there exists $n \in F$ such that $\text{domain}(\varphi_n) = \{\theta(n)+1\}$. (It follows that there is no algorithm which, applied to any computable partial function φ with finite domain, will compute an upper bound for the domain of φ.) Use this to give another solution to the second part of Exercise (5.25.1).

We have already made implicit use of the case $n = 2$ of the following lemma on several occasions.

(5.26) Lemma.   *For each positive integer n there exists a one-one effective enumeration of $\mathbf{N}^n$.*

Proof. This is trivial in the case $n = 1$. Assume, therefore, that there is a one-one total computable function f from $\mathbf{N}$ onto $\mathbf{N}^n$. Following the arrows through the diagram below and deleting all repetitions, we obtain a one-one effective enumeration of $\mathbf{N}^n \times \mathbf{N}$:

$$
\begin{array}{ccccc}
(f(0),0) & \rightarrow & (f(0),1) & & (f(0),2) \quad \rightarrow \quad \cdots \\
& \swarrow & & \nearrow & \\
(f(1),0) & & (f(1),1) & \cdots & \\
\downarrow & \nearrow & & & \\
(f(2),0) & \cdots & & &
\end{array}
$$

$\vdots$

Composing this with the one-one total computable mapping

$$((x_1,\ldots,x_n),m) \mapsto (x_1,\ldots,x_n,m)$$

of $\mathbf{N}^n \times \mathbf{N}$ onto $\mathbf{N}^{n+1}$, we obtain an effective one-one enumeration of $\mathbf{N}^{n+1}$.

$\square$

In spite of Exercise (5.21.3), it is not Proposition (5.18) but the following two theorems, taken together, which are known as the **Extended Version of Rice's Theorem**.

(5.27) Theorem. *There exists a total computable function $f : \mathbf{N} \to \mathbf{N}$ such that for each $m \in \mathbf{N}$,*

(i) $\operatorname{domain}(\varphi_{f(m)}) = \{j \in \mathbf{N} : \textbf{trans}(j) \in \operatorname{domain}(\varphi_m)\}$, *and*

(ii) *if* $\operatorname{domain}(\varphi_m)$ *respects indices, then* $n \in \operatorname{domain}(\varphi_m)$ *if and only if there exists* $k \in \operatorname{domain}(\varphi_{f(m)})$ *such that* $\psi_k \subset \varphi_n$.

Proof. Apply the *s-m-n* theorem to construct a total computable function $f : \mathbf{N} \to \mathbf{N}$ such that $\varphi_{f(m)} = \varphi_m \circ \textbf{trans}$ for each m. Clearly, (i) is satisfied. Consider any $m \in \mathbf{N}$ such that $\operatorname{domain}(\varphi_m)$ respects indices. If $n \in \operatorname{domain}(\varphi_m)$, then by Proposition (5.19), there exists $i \in \operatorname{domain}(\varphi_m)$ such that φ_i is a finite subfunction of φ_n; choosing j such that

$$\varphi_i = \psi_j = \varphi_{\textbf{trans}(j)},$$

we see that as $\operatorname{domain}(\varphi_m)$ respects indices,

$$\textbf{trans}(j) \in \operatorname{domain}(\varphi_m);$$

whence $j \in \operatorname{domain}(\varphi_{f(m)})$. On the other hand, if $j \in \operatorname{domain}(\varphi_{f(m)})$, then $\textbf{trans}(j) \in \operatorname{domain}(\varphi_m)$; if also $\varphi_n \supset \psi_j = \varphi_{\textbf{trans}(j)}$, then Proposition

(5.18) ensures that $n \in \text{domain}(\varphi_m)$. This completes the proof of (ii).

□

(5.28) Theorem.   *Let I be a subset of $\mathbf{N}$, and suppose there is a recursively enumerable subset J of $\mathbf{N}$ such that $i \in I$ if and only if there exists $j \in J$ with $\psi_j \subset \varphi_i$. Then I is recursively enumerable and respects indices.*

Proof.   It is immediate that I respects indices. If J is empty, then so is I, which is therefore recursively enumerable. If $J \neq \emptyset$, then, choosing a total computable function t from $\mathbf{N}$ onto J, for all $m, n, k \in \mathbf{N}$ define

$$
\begin{aligned}
\Psi(m, n, k) \quad &= \quad m \quad \text{if either } d(t(n)) = \emptyset; \text{ or else } d(t(n)) \neq \emptyset \text{ and for} \\
&\qquad\qquad \text{each } j \in d(t(n)), \, \mathcal{M}_m \text{ halts in at most } k+1 \text{ steps} \\
&\qquad\qquad \text{on the input } j, \text{ and } \varphi_m(j) = \psi_{t(n)}(j), \\
&= \quad -1 \quad \text{otherwise.}
\end{aligned}
$$

In view of Lemma (5.23), we see that Ψ is a total computable function from $\mathbf{N}^3$ into $\mathbf{N} \cup \{-1\}$. By Lemma (5.26), there exists a one-one effective enumeration F of $\mathbf{N}^3$. To obtain a recursive enumeration of I, we need only delete all entries equal to -1 from the list $\Psi(F(0)), \Psi(F(1)), \dots$.   □

(5.29) Exercises

.1   Complete the detailed justification of the last sentence in the proof of Theorem (5.28).

.2*   Use Exercise (5.24.1) and Theorem (5.28) to prove that $\{i \in \mathbf{N} : \varphi_i \neq \epsilon\}$ is recursively enumerable.

.3   Let $\theta : \mathbf{N} \to \mathbf{N}$ be a computable partial function whose domain includes $\{i \in \mathbf{N} : \text{domain}(\varphi_i) \text{ is finite}\}$. Prove that there exists n such that (i) $\text{domain}(\varphi_n)$ is nonempty and finite, and (ii) if $\text{domain}(\psi_{\theta(n)})$ is nonempty, then $\psi_{\theta(n)} \not\subset \varphi_n$ (cf. Exercise (5.25.2)).

.4   Prove that the following putative extension of Proposition (5.19) and Theorem (5.27) does not hold: for each recursively enumerable set I that respects indices there exists a total computable function $s : \mathbf{N} \to \mathbf{N}$ such that

(i)   if $n \in I$, then $\psi_{s(n)} = \varphi_k$ for some $k \in I$; and

(ii) $n \in I$ if and only if $\psi_{s(n)} \subset \varphi_n$.

(*Hint*: Use Exercises (5.29.2) and (5.29.3).)

6

Abstract Complexity Theory

So far, we have only concerned ourselves with computability *in principle,* without regard for the efficiency of the computations under discussion. In this chapter we introduce Blum's axiomatic treatment of the theory of the complexity, or cost, of a computation. In practice, this cost is a measure of the amount of some appropriate resource—such as time, space, or memory—used in a computation. The beauty of Blum's axioms is that, in spite of their simplicity and brevity, they enable us to prove a remarkable range of theorems about complexity in the most general context. These theorems hold independently of their interpretation in any model of computation, such as the Turing machine model; our abstract theory of complexity is *machine independent.*

Nevertheless, the basic models for complexity theory are connected with Turing machine computations. For example, we can measure the cost of the computation of $\varphi_i(n)$ by counting the number of steps taken by $\mathcal{M}_i$ to complete a computation (if it does) on the input n. A different measure of the cost is given by the number of distinct cells visited by the read/write head when $\mathcal{M}_i$ computes $\varphi_i(n)$. In either example, if $\varphi_i(n)$ is undefined, we consider the corresponding cost to be undefined.

Following Blum [6], we abstract from these examples a general notion of cost. A **complexity measure** is an infinite sequence

$$\Gamma \equiv \gamma_0, \gamma_1, \gamma_2, \ldots$$

of computable partial functions $\gamma_i : \mathbf{N} \to \mathbf{N}$ that satisfies **Blum's axioms:**

B1 For each i, domain(γ_i) = domain(φ_i).

B2 The function **costs** : $\mathbf{N}^3 \to \{0, 1\}$ defined by

$$\begin{aligned} \mathbf{costs}(i, n, k) &= 1 \quad \text{if } \gamma_i(n) = k, \\ &= 0 \quad \text{otherwise} \end{aligned}$$

is computable.

The computable partial function γ_i is called the **complexity function**, or **cost function**, associated with φ_i.

Blum's axioms are certainly satisfied by the examples of complexity functions described in the second last paragraph. The first of the following exercises proves that the axioms are independent, in the sense that neither can be deduced from the other; the second shows that the axioms are satisfied by some rather unexpected candidates for the title *complexity measure;* and the last three contain elementary results to which we shall refer later.

(6.1) Exercises

.1 Give examples of sequences $\Gamma \equiv \gamma_0, \gamma_1, \gamma_2, \ldots$ of computable partial functions such that

$\qquad\qquad$ (i) Γ satisfies axiom B1 but not B2;

$\qquad\qquad$ (ii) Γ satisfies axiom B2 but not B1.

.2 Let $\Gamma \equiv \gamma_0, \gamma_1, \gamma_2, \ldots$ be a complexity measure, let S be a recursive subset of $\mathbf{N}$, and choose j such that φ_j is the characteristic function of S. Show that

$$\begin{aligned}
\gamma_i' &= 0 \quad \text{if } i = j, \\
&= \gamma_i \quad \text{if } i \neq j
\end{aligned}$$

defines a complexity measure $\Gamma' \equiv \gamma_0', \gamma_1', \gamma_2', \ldots$. Why is it reasonable to call this complexity measure *pathological?* (*Hint*: What if S is the set of all prime numbers?)

.3 Given a complexity measure $\Gamma \equiv \gamma_0, \gamma_1, \gamma_2, \ldots$ and a total computable function $f : \mathbf{N} \to \mathbf{N}$, define

$$\gamma_i' \equiv \gamma_i + f \circ \varphi_i \quad (i \in \mathbf{N}).$$

Prove that $\Gamma' \equiv \gamma_0', \gamma_1', \gamma_2', \ldots$ is a complexity measure.

.4 Given a complexity measure $\Gamma \equiv \gamma_0, \gamma_1, \gamma_2, \ldots$ and a total computable function $t : \mathbf{N} \to \mathbf{N}$, define a total function $G : \mathbf{N}^3 \to \mathbf{N}$ by

$$\begin{aligned}
G(i, n, k) &= 1 \quad \text{if } \gamma_i(n) \leq t(k), \\
&= 0 \quad \text{otherwise.}
\end{aligned}$$

Prove that G is computable.

.5 Given a total computable function $v : \mathbf{N} \to \mathbf{N}$, construct a total computable function $s : \mathbf{N} \to \mathbf{N}$ such that

$$\varphi_{s(i)}(n) = \varphi_{\varphi_i \circ v(n)}(n)$$

whenever either side of this equation is defined. Define a total function $G : \mathbf{N}^4 \to \mathbf{N}$ as follows:

$$\begin{aligned}
G(n, i, j, k) &= \gamma_{s(i)}(n) \quad \text{if } \gamma_i(v(n)) = j \text{ and } \gamma_{\varphi_i \circ v(n)}(n) = k, \\
&= 0 \qquad\qquad \text{otherwise.}
\end{aligned}$$

Prove that G is computable.

From now on, we shall assume that

$$\Gamma \equiv \gamma_0, \gamma_1, \gamma_2, \ldots$$

and

$$\Gamma' \equiv \gamma_0', \gamma_1', \gamma_2', \ldots$$

are complexity measures, and that **costs** (respectively, **costs′**) is the function associated with Γ (respectively, Γ') as in axiom B2.

Our first proposition about complexity shows that the index of a cost function associated with φ_i can be obtained as a computable function of i.

(6.2) Proposition.   *There exists a total computable function $s : \mathbf{N} \to \mathbf{N}$ such that $\gamma_i = \varphi_{s(i)}$ for each i.*

Proof.   In view of the s-m-n theorem, it suffices to observe that the partial function $\Phi : \mathbf{N}^2 \to \mathbf{N}$ defined by $\Phi(i, n) \equiv \gamma_i(n)$ is computable.

$\square$

(6.3) Exercise

> It is clear that the function Φ in the proof of Proposition (6.2) is computable in the case where $\gamma_i(n)$ is the number of steps taken by $\mathcal{M}_i$ in the computation of $\varphi_i(n)$. But why is Φ computable for an arbitrary complexity measure Γ?

Let P be a property applicable to some, but not necessarily all, natural numbers n. We say that $P(n)$ holds **almost everywhere**, or **for almost all values of** n, if there exists ν such that $P(n)$ holds whenever $n \geq \nu$ and P is applicable to n. On the other hand, we say that $P(n)$ holds **infinitely often** if there exist infinitely many values of n such that $P(n)$ holds.

For example, given natural numbers i and j, we say that $\varphi_i(n) \leq \varphi_j(n)$ almost everywhere if there exists ν such that $\varphi_i(n) \leq \varphi_j(n)$ for all $n \geq \nu$ in domain$(\varphi_i) \cap$ domain(φ_j).

The first major result of this section, the **Recursive Relatedness Theorem** for complexity measures, reveals a pleasing symmetry almost everywhere in the expression of recursive bounds for the functions in one complexity measure in terms of their counterparts in another.

(6.4) Theorem.   *For any two complexity measures Γ and Γ' there exists a total computable function $F : \mathbf{N}^2 \to \mathbf{N}$ such that $\gamma_i(n) \leq F(n, \gamma_i'(n))$ and $\gamma_i'(n) \leq F(n, \gamma_i(n))$ for all $n \geq i$ in domain(φ_i).*

Proof.   Define a total function $G : \mathbf{N}^3 \to \mathbf{N}$ as follows:

$$\begin{aligned} G(i, n, k) &= \gamma_i(n) + \gamma_i'(n) && \text{if either } \gamma_i(n) \leq k \text{ or } \gamma_i'(n) \leq k, \\ &= 0 && \text{otherwise.} \end{aligned}$$

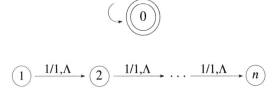

FIGURE 13. The Turing machine $\mathcal{T}_n$.

Note that, by B1, if either $\gamma_i(n)$ or $\gamma_i'(n)$ is defined, then so is the other and hence their sum. On the other hand, Exercise (6.1.4) ensures that G is computable. Therefore

$$F(n, k) \equiv \max_{i \leq n} G(i, n, k)$$

defines a total computable function $F : \mathbf{N}^2 \to \mathbf{N}$. If $n \in \mathrm{domain}(\varphi_i)$ and $n \geq i$, then

$$\gamma_i(n) \leq \gamma_i(n) + \gamma_i'(n) = G(i, n, \gamma_i'(n)) \leq F(n, \gamma_i'(n)).$$

To complete the proof we need only interchange the rôles of Γ and Γ'.

$\square$

There are two obvious ways in which we might seek to improve upon Theorem (6.4): in the first of these we try to remove the *almost everywhere* condition; in the second we try to replace F by a total computable function f of one variable to obtain the inequalities $\gamma_i(n) \leq f(\gamma_i'(n))$ and $\gamma_i'(n) \leq f(\gamma_i(n))$ almost everywhere for each i. We now show that the first of these proposed improvements is impossible; the impossibility of the second is left as Exercise (6.5.2).

Define the complexity measures Γ and Γ' by setting[1]

$$
\begin{aligned}
\gamma_i(n) &\equiv && \text{the number of distinct cells of } \mathcal{M}_i \text{ visited by the} \\
& && \text{read/write head when } \mathcal{M}_i \text{ computes } \varphi_i(n), \\
\gamma_i'(n) &\equiv && \gamma_i(n) + i.
\end{aligned}
$$

Using Exercise (6.1.3), we see that Γ' is a complexity measure. For each natural number n the normalised binary Turing machine $\mathcal{T}_n$ in Figure 13 computes the identity function $\mathbf{id} : \mathbf{N} \to \mathbf{N}$; morever, for each k the number of distinct cells of $\mathcal{T}_n$ visited by the read/write head during the computation of $\mathbf{id}(k)$ is 1.

---

[1]In such definitions it is taken for granted that $\gamma_i(n)$, for example, is undefined if $\mathcal{M}_i$ fails to complete a computation on the input n.

For each n let $h(n)$ be the index of $\mathcal{T}_n$, so that $\mathcal{T}_n = \mathcal{M}_{h(n)}$. Let $F : \mathbf{N}^2 \to \mathbf{N}$ be a total computable function, and choose n so large that $1 + h(n) > F(0,1)$. Setting $i \equiv h(n)$, we have $\gamma_i(0) = 1$ and

$$\gamma_i'(0) = 1 + i > F(0,1) = F(0, \gamma_i(0)).$$

Since F is arbitrary, we conclude that the *almost everywhere* condition cannot be removed from the conclusion of Theorem (6.4).

(6.5) Exercises

.1 Prove that for each complexity measure Γ there exists a complexity measure Γ' with the following property: for each total computable function $F : \mathbf{N}^2 \to \mathbf{N}$ there exists ν such that $\gamma_\nu'(n) > F(n, \gamma_\nu(n))$ for all $n \in \operatorname{domain}(\varphi_\nu)$. (*Hint:* Define $\gamma_i' \equiv 1 + \gamma_i + \varphi_i$.) Can you explain the apparent contradiction between this result and Theorem (6.4)?

.2* With γ_i, γ_i' as in the example preceding this set of exercises, show that for each total computable function $f : \mathbf{N} \to \mathbf{N}$ there exists i such that $\gamma_i'(n) > f(\gamma_i(n))$ for all n. Thus in the conclusion of Theorem (6.4) we cannot replace F by a total computable function $f : \mathbf{N} \to \mathbf{N}$ such that for each i, $\gamma_i(n) \le f(\gamma_i'(n))$ and $\gamma_i'(n) \le f(\gamma_i(n))$ almost everywhere.

Our next result shows that there is a computable bound, independent of i, for the values of φ_i in terms of the values of γ_i.

(6.6) Proposition. *There exists a total computable function $F : \mathbf{N}^2 \to \mathbf{N}$ such that $\varphi_i(n) \le F(n, \gamma_i(n))$ for all i and for all $n \ge i$ in $\operatorname{domain}(\varphi_i)$.*

Proof. Define a total function $H : \mathbf{N}^3 \to \mathbf{N}$ by

$$
\begin{aligned}
H(i, n, k) &= \varphi_i(n) \quad \text{if } \gamma_i(n) = k, \\
&= 0 \qquad\;\; \text{otherwise.}
\end{aligned}
$$

This is computable in view of Blum's axioms. It follows that

$$F(n, k) \equiv \max\{H(i, n, k) : i \le n\}$$

defines a total computable function $F : \mathbf{N}^2 \to \mathbf{N}$. Also, for all i, and all $n \ge i$ in $\operatorname{domain}(\varphi_i)$,

$$F(n, \gamma_i(n)) \ge H(i, n, \gamma_i(n)) = \varphi_i(n). \qquad \square$$

In contrast to Proposition (6.6), there is no computable bound, independent of i, for the values of γ_i in terms of the values of φ_i; that is, there is no total computable function $F : \mathbf{N}^2 \to \mathbf{N}$ such that for each $i, \gamma_i(n) \leq F(n, \varphi_i(n))$ almost everywhere. See Exercise (6.11.2).

The first two of the following exercises reveal the limitations of Proposition (6.6).

(6.7) Exercises

.1   Construct a complexity measure Γ with the following property: for each total computable function $F : \mathbf{N}^2 \to \mathbf{N}$ there exists i such that $\varphi_i(0) > F(0, \gamma_i(0))$. (*Hint:* First construct a one-one total computable function $h : \mathbf{N} \to \mathbf{N}$, with recursive range, such that for each k, $h(k)$ is the index of a Turing machine that computes the constant function $n \mapsto k$.)

.2* Define the complexity measure $\Gamma \equiv \gamma_0, \gamma_1, \ldots$ by

$$\gamma_i(n) \quad \equiv \quad \text{the number of distinct cells of } \mathcal{M}_i \text{ visited by the read/write head when } \mathcal{M}_i \text{ computes } \varphi_i(n).$$

Prove that there exists i such that φ_i is total, and such that for each total computable function $f : \mathbf{N} \to \mathbf{N}$, $\varphi_i(n) > f(\gamma_i(n))$ almost everywhere.

.3* Construct a complexity measure Γ such that $\varphi_i(n) \leq \gamma_i(n)$ for all i and n.

We now have an abstract expression of the common experience that it is possible to construct programs that compute a given partial function and have arbitrarily high cost. To be precise, we show that for each pair (f, t) of total computable functions there exists a computation of f whose cost, at each input, is greater than t. Before doing so, we define the **index set** of a computable partial function $\varphi : \mathbf{N} \to \mathbf{N}$ to be

$$\mathbf{IND}(\varphi) \equiv \{i \in \mathbf{N} : \varphi = \varphi_i\}.$$

(6.8) Proposition.   *Let $t : \mathbf{N} \to \mathbf{N}$ be a total computable function. Then for each total computable function $f : \mathbf{N} \to \mathbf{N}$ there exists $i \in \mathbf{IND}(f)$ such that $\gamma_i(n) > t(n)$ for all n.*

Proof.   Given a total computable function $f : \mathbf{N} \to \mathbf{N}$, define a total function $H : \mathbf{N}^2 \to \mathbf{N}$ as follows:

$$\begin{aligned} H(k, n) &= \varphi_k(n) + 1 & \text{if } \gamma_k(n) \leq t(n), \\ &= f(n) & \text{otherwise.} \end{aligned}$$

In the notation of Exercise (6.1.4) we have

$$H(k,n) \quad = \quad \varphi_k(n) + 1 \quad \text{if } G(k,n,n) = 1,$$
$$= \quad f(n) \qquad\qquad \text{if } G(k,n,n) = 0.$$

Since, by that exercise, G is computable, so is H. By the s-m-n theorem, there exists a total computable function $s : \mathbf{N} \to \mathbf{N}$ such that for each k, $\varphi_{s(k)}$ equals $H(k, \cdot)$ and is therefore total. Applying the Recursion Theorem, we obtain i such that $\varphi_{s(i)} = \varphi_i$. For each n, since $\varphi_i(n) \neq \varphi_i(n) + 1$, the definition of H shows that

$$\varphi_i(n) = \varphi_{s(i)}(n) = H(i,n) = f(n)$$

and that $\gamma_i(n) > t(n)$.   □

The following exercise shows that if $t : \mathbf{N} \to \mathbf{N}$ is a total computable function, then there is a total computable function f such that for *each* Turing machine $\mathcal{M}$ (no matter how well designed) that computes f, the cost of computing $f(n)$ using $\mathcal{M}$ is greater than $t(n)$ for some values of n.

(6.9) Exercise

Construct a total computable function $v : \mathbf{N} \to \mathbf{N}$ such that for each $k \in \mathbf{N}$ there are infinitely many values of n with $v(n) = k$. Given a total computable function $t : \mathbf{N} \to \mathbf{N}$, define a total function $f : \mathbf{N} \to \mathbf{N}$ as follows:

$$f(n) \quad = \quad \varphi_{v(n)}(n) + 1 \quad \text{if } \gamma_{v(n)}(n) \leq t(n),$$
$$= \quad 0 \qquad\qquad\qquad \text{otherwise.}$$

Prove that f is computable and that $\gamma_{v(n)}(n) > t(n)$ whenever $v(n) \in \mathbf{IND}(f)$.

The next result is a strengthening of the preceding exercise.

(6.10) Theorem. *Let $t : \mathbf{N} \to \mathbf{N}$ be a total computable function. Then there exists a total computable function $f : \mathbf{N} \to \{0, 1\}$ such that for each $i \in \mathbf{IND}(f)$, $\gamma_i(n) > t(n)$ almost everywhere.*

Proof. For each pair j, n of natural numbers and each partial function $\varphi : \mathbf{N} \to \mathbf{N}$ define the property P as follows:

$$P(j,n,\varphi) \quad \Leftrightarrow \quad \gamma_j(n) \leq t(n) \text{ and there is no } k < n \text{ such that}$$
$$\gamma_j(k) \leq t(k) \text{ and } \varphi_j(k) \neq \varphi(k).$$

Note that $P(j, n, \varphi_j)$ if and only if $\gamma_j(n) \leq t(n)$. Now define a partial function $\theta : \mathbf{N} \to \mathbf{N}$ and a total function $f : \mathbf{N} \to \{0, 1\}$ simultaneously by the following conditions:

$$
\begin{aligned}
\theta(n) &= 0 & \text{if } n = 0 \text{ and } \gamma_0(0) \leq t(0), \\
&= \text{undefined} & \text{if } n = 0 \text{ and } \neg(\gamma_0(0) \leq t(0)), \\
&= \min\{j \leq n : P(j, n, f)\} & \text{if } n \geq 1,
\end{aligned}
$$

and

$$
\begin{aligned}
f(n) &= 0 \quad \text{if } n \in \operatorname{domain}(\theta) \text{ and } \varphi_{\theta(n)}(n) = 1, \\
&= 1 \quad \text{otherwise.}
\end{aligned}
$$

Then $\operatorname{domain}(\theta)$ is a recursive set, by Exercise (6.1.4), and θ is a computable partial function on $\mathbf{N}$. Moreover, the definition of θ ensures that if $\theta(n)$ is defined, then so are $\gamma_{\theta(n)}(n)$ and (by axiom B1) $\varphi_{\theta(n)}(n)$; in which case we can decide whether $\varphi_{\theta(n)}(n)$ equals 1, so that $f(n)$ is defined. On the other hand, if $n \notin \operatorname{domain}(\theta)$, then $f(n) = 1$, by definition. Hence f is a total computable function on $\mathbf{N}$.

We claim that θ is injective. Indeed, if θ is not injective, then there exist m, n in $\operatorname{domain}(\theta)$ with the following properties: $m < n$, $\theta(m) = \theta(n)$, $\gamma_{\theta(m)}(m) \leq t(m)$, $\gamma_{\theta(n)}(n) \leq t(n)$, and there is no $k < n$ such that $\gamma_{\theta(n)}(k) \leq t(k)$ and $\varphi_{\theta(n)}(k) \neq f(k)$. It follows that

$$
\varphi_{\theta(m)}(m) = \varphi_{\theta(n)}(m) = f(m),
$$

which contradicts the definition of $f(m)$.

Now assume that there exists $i \in \mathbf{IND}(f)$ such that $\gamma_i(n) \leq t(n)$ infinitely often. Then $P(i, n, f)$ holds for infinitely many values of n, since $f = \varphi_i$. For each such n we have $\theta(n) \leq i$, by the definition of θ; but this is absurd, since θ is injective. Hence, in fact, $\gamma_i(n) > t(n)$ for all sufficiently large n.   □

(6.11) Exercises

.1   Define a complexity measure Γ by setting

$$
\gamma_i(n) \equiv \text{the number of steps taken by } \mathcal{M}_i \\
\text{to compute } \varphi_i(n).
$$

Prove that for each $n \in \mathbf{N}$ there exists a total computable function $b : \mathbf{N} \to \mathbf{N}$ with the following property: for each total computable function $f : \mathbf{N} \to \{0, 1\}$ there exists $i \in \mathbf{IND}(f)$ such that $\gamma_i(k) \leq b(k)$ for $k = 0, \ldots, n$. It follows that the *almost everywhere* restriction cannot be removed from the conclusion of Theorem (6.10).

.2* Prove that for each total computable function $F : \mathbf{N}^2 \to \mathbf{N}$ there exists a total computable function $f : \mathbf{N} \to \mathbf{N}$ such that for each $i \in \mathbf{IND}(f)$, $\gamma_i(n) > F(n, \varphi_i(n))$ almost everywhere. (Define $t(n) \equiv F(n,0) + F(n,1)$ and apply (6.10).) This exercise shows that there is no computable bound almost everywhere for the values of the cost function γ_i in terms of the values of the corresponding computable partial function φ_i.

To each total computable function f there corresponds a unique **complexity class** C_f, consisting of all computable partial functions $\varphi : \mathbf{N} \to \mathbf{N}$ with the following property: there exists $i \in \mathbf{IND}(\varphi)$ such that $\gamma_i(n) \leq f(n)$ almost everywhere.[2] Our next aim is to study some basic properties of complexity classes and to prove two fundamental theorems: the gap theorem and the compression theorem.

Define a partial order $\prec$ on the set of complexity classes as follows:

$$C_f \prec C_g \text{ if and only if } C_f \subset C_g \text{ and } C_f \neq C_g.$$

It follows from Theorem (6.10) that for each total computable function f there exists a total computable function g such that $C_f \prec C_g$; thus there exists an infinite ascending sequence $C_{f_0} \prec C_{f_1} \prec C_{f_2} \prec \cdots$ of complexity classes.

(6.12) Exercise

Is the intersection of two complexity classes a complexity class? What about the union of two complexity classes?

If $f : \mathbf{N} \to \mathbf{N}$ and $g : \mathbf{N} \to \mathbf{N}$ are total computable functions and $f(n) \leq g(n)$ almost everywhere, then $C_f \subset C_g$. Can we be sure that if g is much larger than f almost everywhere, then $C_f \prec C_g$? The following **Gap Theorem** will show us that, no matter how much larger than f is g, we may not have $C_f \prec C_g$; indeed, there may be no index i such that $\gamma_i(n) > f(n)$ infinitely often but $\gamma_i(n) \leq g(n)$ almost everywhere.

(6.13) Theorem.   *Let $F : \mathbf{N}^2 \to \mathbf{N}$ be a total computable function such that $F(j,k) > k$ for all j and k. Then for each total computable function $t : \mathbf{N} \to \mathbf{N}$ there exists a total computable function $f : \mathbf{N} \to \mathbf{N}$ such that*

(i) $f(n) \geq t(n)$ *for all n, and*
(ii) *if $f(n) \leq \gamma_i(n) \leq F(n, f(n))$, then $n \leq i$.*

---

[2]Note that for φ to belong to C_f it is not required that *all* algorithms for the computation of φ have complexity bounded by f almost everywhere. Such a requirement would be absurd in view of Proposition (6.8).

Proof.   Let $t : \mathbf{N} \to \mathbf{N}$ be a total computable function. Define a total function $G : \mathbf{N}^2 \to \{0,1\}$ by

$$
\begin{aligned}
G(k,n) \;=\; & 0 \quad \text{if } t(n) \le k \text{ and } \forall i < n \; (\gamma_i(n) < k \\
& \qquad \text{or } \neg(\gamma_i(n) \le F(n,k))), \\
\;=\; & 1 \quad \text{otherwise.}
\end{aligned}
$$

By Exercise (6.1.4), G is computable. (Note that if $\neg(\gamma_i(n) \le F(n,k))$, it does not follow that $\gamma_i(n) > F(n,k)$, as $\gamma_i(n)$ need not be defined.) Given $n \in \mathbf{N}$, we prove that

$$
\mathcal{D}(n) \equiv \{k \in \mathbf{N} : G(k,n) = 0\}
$$

is nonempty. To this end, define

$$
\begin{aligned}
k_j \;=\; & t(n) & \text{if } j = 0, \\
\;=\; & F(n,k_{j-1}) & \text{if } j \ge 1.
\end{aligned}
$$

Then $t(n) = k_0 < k_1 < \cdots$. Since there are at most n values $\gamma_i(n)$ with $0 \le i < n$, there exists r $(0 \le r \le 2n+1)$ such that for all $i < n$, $\gamma_i(n) \notin [k_r, k_{r+1}]$; see Exercise (6.15.1). For this r, either $\gamma_i(n) < k_r$ or $\neg(\gamma_i(n) \le k_{r+1} = F(n,k_r))$, so $k_r \in \mathcal{D}(n)$. It now suffices to let $f : \mathbf{N} \to \mathbf{N}$ be the total function obtained from G by minimization over its first variable,

$$
f(n) \equiv \min k \, [G(k,n) = 0] \quad (n \in \mathbf{N}),
$$

which was proved computable in Exercise (2.7.3).   □

(6.14) Corollary.   *Under the hypotheses of* Theorem (6.13), $C_f = C_{F \circ (P_1^1, f)}$.

Proof.   Since $f(n) \le F(n, f(n))$ for all n,

$$
C_f \subset C_{F \circ (P_1^1, f)}.
$$

On the other hand, if $\varphi_i \in C_{F \circ (P_1^1, f)}$, there exists N such that if $n \ge N$ and $n \in \text{domain}(\varphi_i)$, then $\gamma_i(n) \le F(n, f(n))$. Consider any $n > \max\{i, N\}$ in $\text{domain}(\varphi_i)$. If $f(n) \le \gamma_i(n)$, then $n \le i$, by (6.13); so $\gamma_i(n) < f(n)$. Thus $\gamma_i(n) \le f(n)$ almost everywhere, and so $\varphi_i \in C_f$.   □

To appreciate the force of this corollary, take, for example, $F(m,n) \equiv 2^{2^n}$: there exists a total computable function $f : \mathbf{N} \to \mathbf{N}$ such that if $\gamma_i(n) \le 2^{2^{f(n)}}$ almost everywhere, then $\gamma_i(n) \le f(n)$ almost everywhere!

(6.15) Exercises

.1* Let $k_0 < k_1 < \cdots < k_{2n+2}$ and $c_1, \ldots, c_n$ be real numbers. Prove that there exists r $(0 \le r \le 2n + 1)$ such that $c_j \notin [k_r, k_{r+1}]$ for each j. (This is used in the proof of (6.13).)

.2* Let $s : \mathbf{N} \to \mathbf{N}$ be a total computable function such that $s(n) > n$ for all n. Prove that for each total computable function $t : \mathbf{N} \to \mathbf{N}$ there exists a total computable function $f : \mathbf{N} \to \mathbf{N}$ such that

$$\begin{aligned} &\text{(i)} \quad f(n) \ge t(n) \text{ for all } n, \text{ and} \\ &\text{(ii)} \quad C_f = C_{s \circ f}. \end{aligned}$$

(Several authors call this theorem the *Gap Theorem*.)

.3 Prove the following **Uniform Version of the Gap Theorem**: *Let $F : \mathbf{N}^2 \to \mathbf{N}$ be a total computable function such that $F(j, k) > k$ for all j and k. Then there exists a total computable function $s : \mathbf{N} \to \mathbf{N}$ such that if φ_m is total, then*

$$\begin{aligned} &\text{(i)} \quad \varphi_{s(m)} \text{ is total,} \\ &\text{(ii)} \quad \varphi_m(n) \le \varphi_{s(m)}(n) \text{ for all } n, \text{ and} \\ &\text{(iii)} \quad \text{if } \varphi_{s(m)}(n) \le \gamma_i(n) \le F(n, \varphi_{s(m)}(n)), \text{ then } n \le i. \end{aligned}$$

There is an interesting interpretation of Exercise (6.15.2) in which each step executed by a Turing machine takes one unit of time, and $\gamma_i(n)$ measures the number of steps executed by $\mathcal{M}_i$ in the computation of $\varphi_i(n)$. Thus for each total computable function $f : \mathbf{N} \to \mathbf{N}$, C_f consists of all computable partial functions $\varphi : \mathbf{N} \to \mathbf{N}$ such that for some $i \in \mathbf{IND}(\varphi)$, and for all but finitely many $n \in \text{domain}(\varphi)$, the computation of $\varphi_i(n)$ by $\mathcal{M}_i$ takes time at most $f(n)$. Imagine that because of restrictions on the funds available to us for the purchase of computer time, the only partial functions that we can compute in practice are those which belong to C_f. Now imagine also that the restrictions on our funds are relaxed substantially, so that for some very rapidly growing total computable function $s : \mathbf{N} \to \mathbf{N}$, we can afford to compute any partial function $\varphi : \mathbf{N} \to \mathbf{N}$ that satisfies the following condition: for some $i \in \mathbf{IND}(\varphi)$, and for all but finitely many $n \in \text{domain}(\varphi_i)$, the computation of $\varphi(n)$ by $\mathcal{M}_i$ takes time at most $s(f(n))$. It is natural to hope that if $s(n)$ grows sufficiently rapidly with n, there will be functions that we could not afford to compute under the old funding regime but whose computation is practicable under the new one. Alas, the gap theorem shows that if the initial funding conditions, represented by f, are unfavourable, then the increase of funds, represented by $s \circ f$, will not enable us to compute any additional functions within our still restricted time resources!

(6.16) Exercise

> Let Γ, Γ' be complexity measures such that $\gamma_i'(n) \leq \gamma_i(n)$ for all i and for all $n \in \text{domain}(\varphi_i)$, and let C_f' denote the corresponding complexity class associated with f. Prove that for each total computable function $t : \mathbf{N} \to \mathbf{N}$ there exists a total computable function $f : \mathbf{N} \to \mathbf{N}$ such that (i) $f(n) \geq t(n)$ almost everywhere and (ii) $C_f = C_f'$. (*Hint:* Use Theorems (6.4) and (6.13).)

The last exercise has an interpretation in which we consider $\gamma_i(n)$ and $\gamma_i'(n)$ to be the respective times required to compute $\varphi_i(n)$ by implementing simulations of $\mathcal{M}_i$ on two different computers C and C': however fast the processor of C' is, compared with that of C, there exist arbitrarily large total computable functions whose associated complexity classes with respect to C and C' are the same. In other words, increasing the speed of a processor will not necessarily augment the class of partial functions that can be computed within a given time.

If we consider the complexity class of a cost function γ_i, then the picture presented by the gap theorem is totally transformed: a consequence of the following **Compression Theorem** is that there exists a total computable function F of two variables such that if the complexity bound is increased from γ_i to $F(\cdot, \gamma_i(\cdot))$, then the class of partial functions that can be computed in practice is enlarged to include a computable partial function that is relatively small (in that it is bounded by the identity function throughout its domain) and for which each algorithm costs more almost everywhere than γ_i.

(6.17) Theorem.   *There exist total computable functions $s : \mathbf{N} \to \mathbf{N}$ and $F : \mathbf{N}^2 \to \mathbf{N}$ such that for each i,*

(i) $\text{domain}(\varphi_i) = \text{domain}(\varphi_{s(i)})$;
(ii) $\varphi_{s(i)}(n) \leq n$ *for each* $n \in \text{domain}(\varphi_i)$;
(iii)   *for each* $j \in \mathbf{IND}(\varphi_{s(i)})$, $\gamma_j(n) > \gamma_i(n)$ *almost everywhere;*
(iv) $\gamma_{s(i)}(n) \leq F(n, \gamma_i(n))$ *almost everywhere.*

Proof.   Define predicates P and Q by

$$P(i, j, n) \iff j < n \text{ and } \gamma_j(n) \leq \gamma_i(n),$$
$$Q(i, k, n) \iff k \neq \varphi_j(n) \text{ for all } j \text{ such that } P(i, j, n).$$

You are invited, in Exercise (6.19), to show that

$$\Psi(i, n) = \min\{k : Q(i, k, n)\} \quad \text{if } \varphi_i(n) \text{ is defined,}$$
$$= \text{undefined} \qquad\qquad \text{otherwise}$$

defines a computable partial function on $\mathbf{N}^2$, and that if $\Psi(i, n)$ is defined, then $\Psi(i, n) \leq n$. Using the *s-m-n* theorem, construct a total computable function $s : \mathbf{N} \to \mathbf{N}$ such that $\varphi_{s(i)} = \Psi(i, \cdot)$ for each i. Clearly, both (i) and (ii) hold.

Let $j \in \mathbf{IND}(\varphi_{s(i)})$, and consider any $n > j$ such that

$$\varphi_j(n) = \varphi_{s(i)}(n) = \Psi(i, n)$$

is defined. The definition of Ψ ensures that $Q(i, \varphi_j(n), n)$. It follows from the definition of Q that $\neg P(i, j, n)$; whence $\gamma_j(n) > \gamma_i(n)$. This proves (iii).

For all i, n, and k, axiom B2 enables us to decide whether or not $\gamma_i(n)$ equals k; if $\gamma_i(n) = k$, then, by axiom B1, $\varphi_i(n)$ is defined; whence, by the foregoing and B1, both $\varphi_{s(i)}(n)$ and are defined. Thus

$$
\begin{aligned}
H(i, n, k) &= \gamma_{s(i)}(n) \quad \text{if } \gamma_i(n) = k, \\
&= 0 \qquad\qquad \text{otherwise}
\end{aligned}
$$

defines a total computable function $H : \mathbf{N}^3 \to \mathbf{N}$. Now define a total computable function $F : \mathbf{N}^2 \to \mathbf{N}$ by

$$F(n, k) \equiv \max\{H(i, n, k) : i \le n\}.$$

Consider any i and n with $n \ge i$. From the foregoing we see that $\gamma_i(n)$ is defined if and only if $\gamma_{s(i)}(n)$ is defined; in which case,

$$\gamma_{s(i)}(n) = H(i, n, \gamma_i(n)) \le F(n, \gamma_i(n)).$$

Thus (iv) obtains, and our proof is complete.   □

(6.18) Corollary.   *Under the hypotheses of* Theorem (6.17),

$$C_{\gamma_i} \subset C_{Fo(P_1^1, \gamma_i)} \text{ and } C_{\gamma_i} \ne C_{Fo(P_1^1, \gamma_i)}$$

for each $i \in \mathbf{N}$.

Proof.   It follows from (iii) and (iv) of Theorem (6.17) that for each i,

$$C_{\gamma_i} \subset C_{Fo(P_1^1, \gamma_i)}$$

and

$$\varphi_{s(i)} \in C_{Fo(P_1^1, \gamma_i)} \setminus C_{\gamma_i}.   □$$

(6.19) Exercise

Show that, in the proof of (6.17), the function Ψ is computable, and that if $\Psi(i, n)$ is defined, then $\Psi(i, n) \le n$.

We end this chapter by discussing the most famous and startling result of Blum's complexity theory:

(6.20) The Speed-up Theorem [7]. *If* $F : \mathbf{N}^2 \to \mathbf{N}$ *is a total computable function such that* $F(m, n) \leq F(m, n + 1)$ *for all* m *and* n, *then there exists a total computable function* $f : \mathbf{N} \to \mathbf{N}$ *that satisfies* $f(n) \leq n$ *for each* n *and that has the following* **speed-up property**: *for each* $i \in \mathbf{IND}(f)$ *there exists* $j \in \mathbf{IND}(f)$ *such that* $F(n, \gamma_j(n)) \leq \gamma_i(n)$ *almost everywhere.*

In interpreting this theorem we will find it helpful to think of $F(m, n)$ as a very rapidly growing function of its second argument n, and of $\gamma_i(n)$, when defined, as measuring the time taken by $\mathcal{M}_i$ to compute $\varphi_i(n)$. The theorem then says that there exists a total computable function f of one variable, such that, whatever Turing machine $\mathcal{M}_i$ we choose to compute f, there always exists a Turing machine $\mathcal{M}_j$ that computes f and *almost everywhere* does so in a time $\gamma_j(n)$ that satisfies $F(n, \gamma_j(n)) \leq \gamma_i(n)$. So there is no algorithm that, for this particular function f and infinitely many values of n, computes the value $f(n)$ more rapidly than any other algorithm.

For example, taking $F(m, n) \equiv 2^n$, we obtain a total computable function $f : \mathbf{N} \to \mathbf{N}$ and a sequence $i_0, i_1, \ldots$ in $\mathbf{IND}(f)$, such that for each k, $\gamma_{i_{k+1}}(n) \leq \log_2 \gamma_{i_k}(n)$ almost everywhere. Thus

$$
\begin{aligned}
\gamma_{i_1}(n) &\leq \log_2 \gamma_{i_0}(n) \quad \text{almost everywhere,} \\
\gamma_{i_2}(n) &\leq \log_2 \log_2 \gamma_{i_0}(n) \quad \text{almost everywhere,} \\
\gamma_{i_3}(n) &\leq \log_2 \log_2 \log_2 \gamma_{i_0}(n) \quad \text{almost everywhere,}
\end{aligned}
$$

and so on.

There is another way of looking at the Speed-up Theorem. Consider, for example, two computers, one of which runs a million times as fast as the other. Applying the Speed-up Theorem with $F(m, n) \equiv 10^6 n$, we obtain a total computable function f with the following property: to each program P that computes f there corresponds another program P' that computes f, such that for almost all values of n it is at least as quick to compute $f(n)$ using program P' on the slower computer as using program P on the faster one.

The proof of the Speed-up Theorem depends on the following preliminary result known as the **Pseudo-speed-up Theorem.**

(6.21) Theorem. *Let* $F : \mathbf{N}^2 \to \mathbf{N}$ *be a total computable function, and fix a total computable function* $s : \mathbf{N} \to \mathbf{N}$ *such that*

$$
\varphi_{s(i,j)} = \varphi_i^{(2)}(j, \cdot) \quad (i, j \in \mathbf{N}).
$$

Then there exists an index e such that, with

$$f \equiv \varphi_e^{(2)}(0, \cdot), \qquad f_i \equiv \varphi_e^{(2)}(i, \cdot) \quad (i \in \mathbf{N}), \tag{6.1}$$

the following properties hold:

(i) $\varphi_e^{(2)}$ *is a total function on* $\mathbf{N}^2$, *and* $\varphi_e^{(2)}(i, n) \leq n$ *for all* i, n;

(ii)   *for each* i, $f_i = f$ *almost everywhere;*

(iii)   *if* $i \in \mathbf{IND}(f)$, *then* $s(e, i+1) \in \mathbf{IND}(f_{i+1})$, *and*

$$F(n, \gamma_{s(e,i+1)}(n)) \leq \gamma_i(n) \text{ for all } n > i.$$

Proof. To begin with, we define a partial function $\mathcal{C}$ from $\mathbf{N}^3$ into $\mathcal{P}(\mathbf{N})$ (the power set of $\mathbf{N}$) as follows. For each $(e, i, n) \in \mathbf{N}^3$, if $i \geq n$, then $\mathcal{C}(e, i, n)$ is defined to be $\emptyset$. If $i < n$, if $\mathcal{C}(e, i, m)$ is defined whenever $0 \leq m < n$, and if $\gamma_{s(e,j+1)}(n)$ is defined whenever $i \leq j < n$, then

$$\mathcal{C}(e, i, n) \equiv \{j \in \mathbf{N} : i \leq j < n, \ j \notin \bigcup_{m=0}^{n-1} \mathcal{C}(e, i, m),$$
$$\text{and } \gamma_j(n) < F(n, \gamma_{s(e,j+1)}(n))\}.$$

Otherwise, $\mathcal{C}(e, i, n)$ is undefined.[3] We note the following facts about $\mathcal{C}$ whose proofs are left to Exercises (6.22):

- If $\mathcal{C}(e, i, n)$ is defined, then it is a finite recursive set.

- If $n \geq 0$, and $\mathcal{C}(e, i, n)$ is defined, then

$$\mathcal{C}(e, i, n) = \mathcal{C}(e, 0, n) \cap \{i, i+1, \ldots, n-1\}.$$

If $\mathcal{C}(e, i, n)$ is defined and empty, set $\Phi(e, i, n) \equiv 0$; if $\mathcal{C}(e, i, n)$ is defined and nonempty, set

$$\Phi(e, i, n) \equiv \min\{m : \forall j \in \mathcal{C}(e, i, n) \ (m \neq \varphi_j(n))\};$$

otherwise, $\Phi(e, i, n)$ undefined. We prove that the partial function $\Phi : \mathbf{N}^3 \to \mathbf{N}$ is computable. Clearly, in seeking to compute $\Phi(e, i, n)$ we may assume that $0 \leq i < n$. Noting that $\Phi(e, i, 0) = 0$, and that if $\Phi(e, i, n)$ is defined, then so are the values $\Phi(e, i, 0), \ldots, \Phi(e, i, n-1)$, suppose that these values are defined and have been computed. We compute $\Phi(e, i, n)$ by following these instructions:

▷ Run the Turing machines $\mathcal{M}_{s(e,j+1)}$ $(i \leq j < n)$ simultaneously on the input n.

---

[3] A common name for $\mathcal{C}(e, i, n)$ in the literature of complexity theory is **the set of indices cancelled at stage** n **by** e **and** i.

▷ If each of these computations is completed, then use Exercise (6.1.4) to check whether

$$\gamma_j(n) < F(n, \gamma_{s(e,j+1)}(n)) \quad (i \leq j < n). \tag{6.2}$$

▷ If condition (6.2) is not satisfied, then $\Phi(e, i, n)$ is undefined.

▷ If condition (6.2) is satisfied, then from those j such that $i \leq j < n$ and $\gamma_j(n) < F(n, \gamma_{s(e,j+1)}(n))$, select those that do not belong to $\bigcup_{m=0}^{n-1} \mathcal{C}(e, i, m)$. (This selection can be done algorithmically, since $\mathcal{C}(e, i, m)$ is a recursive set for $0 \leq m < n$.)

▷ If there are no such j, then output $\Phi(e, i, n) = 0$. Otherwise, compute $\varphi_j(n)$ for each of the selected j, compute the least natural number m distinct from each of those values $\varphi_j(n)$, and output $\Phi(e, i, n) = m$. (For each selected j, $\varphi_j(n)$ is defined in view of axiom B1.)

If $\Phi(e, i, n)$ is defined, then it is at most n, since there are at most n values $\varphi_j(n)$ with $j \in \mathcal{C}(e, i, n)$.

Using the s-m-n theorem and the Recursion Theorem, we now fix the value of e as one satisfying

$$\Phi(e, \cdot, \cdot) = \varphi_e^{(2)}.$$

The foregoing shows that $\varphi_e^{(2)}(i, n) \leq n$ for all i, n. Define f and f_i $(i \geq 0)$ as in equation (6.1). In proving that f is total we actually prove that for each n,

$$\varphi_e^{(2)}(i, m) \text{ is defined for all } i \text{ and for } 0 \leq m \leq n. \tag{6.3}$$

This is trivially true for $n = 0$: in fact, $\varphi_e^{(2)}(i, 0) = 0$, as $\mathcal{C}(e, i, 0)$ is defined but empty. Assume, for the purpose of induction, that condition (6.3) holds for all $n < \nu$, and consider the case $n = \nu$. If $i \geq \nu$, then $\mathcal{C}(e, i, \nu)$ is defined and empty, so $\varphi_e^{(2)}(i, \nu)$ is defined and equals 0. If $i < \nu$, then, noting that $\varphi_e^{(2)}(\nu, \nu)$ is defined, suppose that

$$\varphi_e^{(2)}(k + 1, \nu) = \varphi_{s(e,k+1)}(\nu) \text{ is defined for } i \leq k < \nu.$$

Axiom B1 then ensures that $\gamma_{s(e,k+1)}(\nu)$ is defined for $i \leq k < \nu$. Since, as readily follows from our original induction hypothesis, $\mathcal{C}(e, i, m)$ is defined for $0 \leq m \leq \nu - 1$, we now see that $\mathcal{C}(e, i, \nu)$ is defined; whence $\Phi(e, i, \nu)$— that is, $\varphi_e^{(2)}(i, \nu)$—is defined. This completes the (forwards and backwards) inductive proof that $\varphi_e^{(2)} : \mathbf{N}^2 \to \mathbf{N}$ is total.

It now follows that $\mathcal{C}(e, i, n)$ is defined, and therefore a finite recursive set, for all i, n. Since, by construction,

$$\mathcal{C}(e, 0, m) \cap \mathcal{C}(e, 0, n) = \emptyset \quad (0 \leq m < n),$$

for each j there exists at most one n such that $j \in C(e, 0, n)$. Hence for each i there exists[4] N_i such that if $0 \le j < i$ and $j \in C(e, 0, n)$, then $n \le N_i$. So if $n > N_i$, then

$$C(e, 0, n) \subset \{i, i+1, \ldots, n-1\}$$

and therefore

$$C(e, i, n) = C(e, 0, n) \cap \{i, i+1, \ldots, n-1\} = C(e, 0, n).$$

It follows from our choice of e, and the relevant definitions, that

$$f_i(n) = \Phi(e, i, n) = \Phi(e, 0, n) = f_0(n) \quad (n > N_i),$$

so conclusion (ii) of Theorem (6.21) holds.

It remains to prove (iii). Given an index i for f, we have

$$\varphi_{s(e,i+1)} = \varphi_e^{(2)}(i+1, \cdot) = f_{i+1},$$

so $s(e, i+1) \in \mathbf{IND}(f_{i+1})$. Suppose that $F(n, \gamma_{s(e,i+1)}(n)) > \gamma_i(n)$ for some $n > i$, and consider the least such value of n. If $i \in C(e, 0, m)$ for some $m < n$, then, by the definition of $C(e, 0, m)$, $i < m < n$ and $F(n, \gamma_{s(e,i+1)}(m)) > \gamma_i(m)$, which contradicts our choice of n. Thus

$$i \notin C(e, 0, m) \quad (m < n).$$

Our choice of n now ensures that $i \in C(e, 0, n)$; whence

$$f(n) = \varphi_e^{(2)}(0, n) = \Phi(e, 0, n) \ne \varphi_i(n),$$

by the definition of Φ. This contradicts our choice of i as an index of f, so we must have

$$F(n, \gamma_{s(e,i+1)}(n)) \le \gamma_i(n) \quad (n > i).$$

Our proof of the Pseudo-speed-up Theorem is now complete. □

(6.22) Exercises

.1 Prove that if $C(e, i, n)$ is defined, then it is a finite recursive set. (*Hint*: Use induction on n.)

.2* Prove that if $n \ge 0$, and $C(e, i, n)$ is defined, then

$$C(e, i, n) = C(e, 0, n) \cap \{i, i+1, \ldots, n-1\}.$$

---

[4] As will be shown below, we may not be able to *compute* this value N_i.

(6.23) Corollary.   *Let F and s be as in* Theorem (6.21). *Then the index e can be chosen so that all the conclusions of* Theorem (6.21) *hold, except that* (iii) *is replaced by the following: if $i \in \mathbf{IND}(f)$, then $s(e, i+1) \in \mathbf{IND}(f_{i+1})$, and*

$$F(n, \gamma_{s(e,i+1)}(n) + n) \leq \gamma_i(n) \text{ for all } n > i.$$

Proof.   It suffices to show that

$$\gamma_i'(n) \equiv \gamma_i(n) + n$$

defines a complexity measure $\Gamma' \equiv \gamma_0', \gamma_1', \dots$, and then to replace γ_i by γ_i' in Theorem (6.21). The details are left as an exercise (cf. Exercise (6.1.3)). □

We now turn to the long-awaited

Proof of the Speed-up Theorem.   We first prove the theorem in the special case where the complexity measure is $\gamma_0^*, \gamma_1^*, \dots$, defined as follows:

$$\gamma_i^*(n) \quad \equiv \quad \text{the number of distinct cells of } \mathcal{M}_i \text{ visited by the}$$
$$\text{read/write head when } \mathcal{M}_i \text{ computes } \varphi_i(n).$$

With f as in Corollary (6.23), fix a normalised binary Turing machine $\mathcal{T}$ that computes f. Consider any $i \in \mathbf{IND}(f)$ and set $k \equiv s(e, i+1)$. There exists N such that $\varphi_k(n) = f(n)$ for all $n \geq N$. We modify $\mathcal{M}_k$ as follows, to produce a normalised binary Turing machine $\mathcal{M}$ that computes f. If, in its start state, $\mathcal{M}$ is given the input $n \in \mathbf{N}$, it first checks to see whether $n < N$. If that is the case, $\mathcal{M}$ calls $\mathcal{T}$ as a module and completes a computation with output $f(n)$; otherwise, $\mathcal{M}$ calls $\mathcal{M}_k$ as a submodule and again completes a computation with output $f(n)$. Let j be the index of $\mathcal{M}$. Clearly, we can arrange the construction of $\mathcal{M}$ so that there exists a constant $c > 0$, depending on k and n, such that

$$\gamma_j^*(n) \leq \gamma_k^*(n) + c \quad (n \in \mathbf{N}).$$

We assume that this has been done. Since F is an increasing function of its second variable, for $n \geq c$ we have

$$F(n, \gamma_j^*(n)) \leq F(n, \gamma_k^*(n) + c) \leq F(n, \gamma_k^*(n) + n).$$

But $F(n, \gamma_k^*(n) + n) \leq \gamma_i^*(n)$ almost everywhere, by Corollary (6.23). Hence $F(n, \gamma_j^*(n)) \leq \gamma_i^*(n)$ almost everywhere, so the proof of our special case of the Speed-up Theorem is complete.

We use the Recursive Relatedness Theorem (6.4) to prove the general case. Accordingly, let $\gamma_0, \gamma_1, \ldots$ be any complexity measure, $\gamma_0^*, \gamma_1^*, \ldots$ the particular complexity measure defined above, and $G : \mathbf{N}^2 \to \mathbf{N}$ a total computable function such that

$$\gamma_i^*(n) \leq G(n, \gamma_i(n)) \text{ and } \gamma_i(n) \leq G(n, \gamma_i^*(n))$$

whenever $n \in \text{domain}(\varphi_i)$ and $n \geq i$. Replacing $G(m, n)$ by

$$n + \max_{0 \leq k \leq n} G(m, k)$$

if necessary, we may assume that $G(m, n) < G(m, n+1)$ for all m and n. Now let f be the function produced by our special case of the theorem, with F replaced by the function

$$(m, n) \mapsto G(m, F(m, G(m, n))).$$

Given an index i of f, we obtain j such that $\varphi_j = f$ and

$$G(n, F(n, G(n, \gamma_j^*(n)))) \leq \gamma_i^*(n) \text{ almost everywhere.}$$

Thus *almost everywhere* we have

$$\begin{aligned} G(n, F(n, \gamma_j(n))) &\leq G(n, F(n, G(n, \gamma_j^*(n)))) \\ &\leq \gamma_i^*(n) \\ &\leq G(n, \gamma_i(n)). \end{aligned}$$

Since G is a *strictly* increasing function of its second argument, it follows that $F(n, \gamma_j(n)) \leq \gamma_i(n)$ almost everywhere. This completes the proof of the general case of the Speed-up Theorem.   □

Let $F : \mathbf{N}^2 \to \mathbf{N}$ and $f : \mathbf{N} \to \mathbf{N}$ be total computable functions. We say that f is F-**speedable** (relative to the complexity measure Γ) if for each $i \in \mathbf{IND}(f)$ there exists $j \in \mathbf{IND}(f)$ such that $F(n, \gamma_j(n)) \leq \gamma_i(n)$ almost everywhere. The Speed-up Theorem says that if F is an increasing function of its second argument, then there exist F-speedable functions.

(6.24) Exercises

.1   Give an example of a total computable function $f : \mathbf{N} \to \mathbf{N}$ and a complexity measure Γ with the following property: if $F : \mathbf{N}^2 \to \mathbf{N}$ is a total computable function such that $F(m, n+1) \geq F(m, n)$ for all m and n, then f is not F-speedable relative to Γ.

.2*   Show that under the hypotheses of the Speed-up Theorem, the conclusion holds for some total computable function f that assumes only the values 0 and 1.

.3 Show that the *almost everywhere* condition cannot be removed from the conclusion of the Speed-up Theorem.

Our proof of the Pseudo-speed-up Theorem is constructive, in that it shows how to compute both e and the index $s(e, i{+}1)$ of the Turing machine that, relative to $\mathcal{M}_i$, speeds up the computation of f. But our proof of the full Speed-up Theorem is not constructive: in the special case where $\gamma_i = \gamma_i^*$ it does not tell us how to find the index j of the faster Turing machine. This defect is not confined to that particular proof. Blum [7] has proved that if $F : \mathbf{N}^2 \to \mathbf{N}$ is a total computable function, and $f : \mathbf{N} \to \mathbf{N}$ an F-speedable total computable function, then there is no computable partial function $\tau : \mathbf{N} \to \mathbf{N}$ such that for each $i \in \mathbf{IND}(f)$, $\tau(i)$ is defined, $\tau(i) \in \mathbf{IND}(f)$, and $F(n, \gamma_{\tau(i)}(n)) \leq \gamma_i(n)$ almost everywhere. The next lemma will enable us to obtain a weak version of this result.

(6.25) Lemma. *Let $v : \mathbf{N} \to \mathbf{N}$ and $s : \mathbf{N} \to \mathbf{N}$ be total computable functions such that*

$$\varphi_{s(i)}(n) = \varphi_{\varphi_i \circ v(n)}(n)$$

whenever either side of this equation is defined. Then there exists a total computable function $H : \mathbf{N}^2 \to \mathbf{N}$ such that

$$\gamma_{s(i)}(n) \leq H(n, \gamma_{\varphi_i \circ v(n)}(n)) + H(i, \gamma_i(v(n)))$$

whenever both sides of this inequality are defined.

Proof. Define the total computable function $G : \mathbf{N}^4 \to \mathbf{N}$ as in Exercise (6.1.5). Define also total computable functions $T : \mathbf{N}^2 \to \mathbf{N}$ and $H : \mathbf{N}^2 \to \mathbf{N}$ by

$$
\begin{aligned}
T(p, q) &\equiv \max\{G(n, i, j, k) : n, k \leq p; \ i, j \leq q\}, \\
H(m, n) &\equiv \max\{T(p, q) : p, q \leq \max\{m, n\}\}.
\end{aligned}
$$

A routine calculation shows that for all n, i, j, k we have

$$
\begin{aligned}
G(n, i, j, k) &\leq T(\max\{n, k\}, \max\{i, j\}) \\
&\leq H(n, k) + H(i, j).
\end{aligned}
$$

Since, by the definition of G,

$$\gamma_{s(i)}(n) = G(n, i, \gamma_i(v(n)), \gamma_{\varphi_i \circ v(n)}(n)),$$

we complete the proof by taking $j \equiv \gamma_i(v(n))$ and $k \equiv \gamma_{\varphi_i \circ v(n)}(n)$. $\square$

(6.26) Proposition. *There exists a total computable function $F : \mathbf{N}^2 \to \mathbf{N}$, increasing in its second argument, with the following property: for each F-speedable function f there is no total computable function $t : \mathbf{N} \to \mathbf{N}$ such that*

(i) $t(i) \in \mathbf{IND}(f)$ *for each* $i \in \mathbf{N}$, *and*

(ii) *for each* $i \in \mathbf{IND}(f)$, $F(n, \gamma_{t(i)}(n)) \leq \gamma_i(n)$ *almost everywhere.*

Proof.   Let $v : \mathbf{N} \to \mathbf{N}$ be a total computable function such that for each $i \in \mathbf{N}$ there exist infinitely many n with $v(n) = i$ (cf. Exercise (6.9)). By the *s-m-n* theorem, there exists a total computable function $s : \mathbf{N} \to \mathbf{N}$ such that

$$\varphi_{s(i)}(n) = \varphi_{\varphi_i \circ v(n)}(n)$$

whenever either side of this equation is defined. Construct H as in Lemma (6.25), and choose a total computable function $F : \mathbf{N}^2 \to \mathbf{N}$ such that

$$F(n, k+1) \geq F(n, k) > n + H(n, k) \quad (n, k \in \mathbf{N}). \tag{6.4}$$

The Speed-up Theorem guarantees the existence of F-speedable functions. Let $f : \mathbf{N} \to \mathbf{N}$ be any one of those, and suppose there exists a total computable function $t : \mathbf{N} \to \mathbf{N}$ with the desired properties relative to F and f. Choose an index k for t. Then

$$\varphi_{s(k)}(n) = \varphi_{\varphi_k \circ v(n)}(n) = \varphi_{t \circ v(n)}(n) = f(n) \quad (n \in \mathbf{N}),$$

so $s(k) \in \mathbf{IND}(f)$. On the other hand, since

$$\gamma_{s(k)}(n) \leq H(n, \gamma_{t \circ v(n)}(n)) + H(k, \gamma_k(v(n))) \quad (n \in \mathbf{N}),$$

we see from our choice of v that there exist infinitely many values of n such that

$$\gamma_{s(k)}(n) \leq H(n, \gamma_{t \circ s(k)}(n)) + H(k, \gamma_k(s(k))).$$

So for infinitely many sufficiently large values of n we have

$$
\begin{aligned}
\gamma_{s(k)}(n) \quad &\leq \quad F(n, \gamma_{t \circ s(k)}(n)) - n + H(k, \gamma_k(s(k))) \\
&\qquad\qquad \text{(by inequality (6.3))} \\
&< \quad F(n, \gamma_{t \circ s(k)}(n)).
\end{aligned}
$$

Since $s(k) \in \mathbf{IND}(f)$, we have arrived at a contradiction.   □

There is another sense in which the Speed-up Theorem is not constructive. If F grows rapidly enough, there is no total computable function $b : \mathbf{N} \to \mathbf{N}$ with the following property: for each F-speedable binary function f and each index i of f there exists an index j of f such that $F(n, \gamma_j(n)) \leq \gamma_i(n)$ whenever $n \geq b(i)$. In other words, there is no computable bound $b(i)$ for the exceptional values of n in the speed-up of f. This result is a simple consequence of the first exercise in the next set.[5]

---

[5]Despite a number of references to this result in the literature, its proof does not appear to be published anywhere (although a related result is proved in [30]). A proof of the full theorem, and the solutions to Exercises (6.27), will be published as "On recursive bounds for the exceptional values in speed-up", by Cristian Calude and the author, in *Theoretical Computer Science.*

(6.27) Exercises

.1* Prove that there exists a total computable function $B : \mathbf{N} \to \mathbf{N}$ with the following property: if $F : \mathbf{N}^2 \to \mathbf{N}$ is a total computable function, increasing in its second argument, such that $F(n, 0) > B(n)$ for each n, if $s \in \mathbf{N}$, and if $f : \mathbf{N} \to \{0, 1\}$ is an F-speedable function such that $\varphi_s(i)$ is defined for each $i \in \mathbf{IND}(f)$, then there exist an index k of f, and a natural number $m > \varphi_s(k)$, such that $\gamma_k(m) < F(m, \gamma_j(m))$ for all $j \in \mathbf{IND}(f)$. (*Hint*: Define a computable partial function $E : \mathbf{N}^5 \to \mathbf{N}$ by

$$E(u, v, i, z, s) \equiv 1 + \max\{i, z, s, \varphi_s(u), \varphi_s(v), \gamma_s(u), \gamma_s(v)\},$$

and use the Double Recursion Theorem (Exercise (5.14.12)) to obtain total computable functions $g_t : \mathbf{N}^3 \to \mathbf{N}$ $(t = 0, 1)$ such that

$$\begin{aligned} \varphi_{g_t(i,z,s)}(n) &= t && \text{if } n = E(g_0(i, z, s), g_1(i, z, s), i, z, s), \\ &= \varphi_i(n) && \text{otherwise.} \end{aligned}$$

Then define

$$\begin{aligned} B(n) \equiv{}& 1 + \max\{\gamma_{g_t(i,z,s)}(n) : t = 0, 1; \ i, z, s \in \mathbf{N}; \\ & E(g_0(i, z, s), g_1(i, z, s), i, z, s) = n\}.) \end{aligned}$$

.2* Prove that if $F : \mathbf{N}^2 \to \mathbf{N}$ is a total computable function, increasing in its second argument, then for each F-speedable function f and each index i of f there exists an index j of f such that $F(n, \gamma_j(n)) \leq \gamma_i(n)$ whenever $n > j$. (*Hint*: First consider the acceptable programming system

$$\varphi_0, \varphi_1, \varphi_0, \varphi_0, \varphi_1, \varphi_1, \varphi_2, \varphi_0, \varphi_0, \varphi_0, \varphi_1, \varphi_1, \varphi_1, \varphi_2, \varphi_2, \varphi_3, \cdots$$

with corresponding complexity measure

$$\gamma_0, \gamma_1, \gamma_0, \gamma_0, \gamma_1, \gamma_1, \gamma_2, \gamma_0, \gamma_0, \gamma_0, \gamma_1, \gamma_1, \gamma_1, \gamma_2, \gamma_2, \gamma_3, \cdots)$$

.3* With B as in Exercise (6.27.1), let $F : \mathbf{N}^2 \to \mathbf{N}$ be a total computable function, increasing in its second argument, such that $F(n, 0) > B(n)$ for each n, and let $f : \mathbf{N} \to \{0, 1\}$ be an F-speedable function. Prove that there is no computable partial function $\theta : \mathbf{N} \to \mathbf{N}$ with the following property: for each index i of f, $\theta(i)$ is defined and there exists $j \leq \theta(i)$ such that (i) $\varphi_j = f$ and (ii) $F(n, \gamma_j(n)) \leq \gamma_i(n)$ for all $n \geq \theta(i)$. (This result should be compared with the theorem of Schnorr [30].)

As the complicated, even pathological, construction used to prove the Pseudo-speed-up Theorem suggests, speedable functions are hard to find. Does this mean that the Speed-up Theorem, fascinating though it may be in theory, is devoid of practical significance? Not necessarily: for a given total computable function $F : \mathbf{N}^2 \to \mathbf{N}$, increasing in its second argument, the F-speedable functions form a set that, in a Baire categorical sense, is much larger than its complement in the set of all total computable functions from $\mathbf{N}$ to $\mathbf{N}$ [10]; so speedable functions are much commoner than those that are not speedable!

The relationship between the set of F-speedable functions and its complement is similar to that between the set of irrational numbers and $\mathbf{Q}$: although, to the numerically naive, irrational numbers seem thinner on the ground than rationals, the set of irrational numbers has larger cardinality than $\mathbf{Q}$; this is clearly shown by the fact that $\mathbf{Q} \cap [0,1]$ has Lebesgue measure 0, whereas the irrational numbers in $[0,1]$ form a set of Lebesgue measure 1.

Epilogue

...all experience is an arch wherethrough
Gleams that untravelled world, whose margin fades
For ever and for ever when I move.
How dull it is to pause, to make an end,
To rust unburnished, not to shine in use!
As though to breathe were life!

ALFRED, LORD TENNYSON, *Ulysses*

Solutions to Exercises

And suppose we solve all the problems...? What happens? We end up with more problems than we started with. Because that's the way problems propagate their species. A problem left to itself dries up or goes rotten. But fertilize a problem with a solution— you'll hatch out dozens.

N.F. SIMPSON, *A Resounding Tinkle,* Act I, Sc. I

Solutions for Chapter 1

(1.2.1)   *Yes.* Examining the sequence of configurations followed by $\mathcal{M}$ when the initial configuration is (Λ, q_0, v), we can determine the rightmost cell c visited by $\mathcal{M}$ before it halts. We can then check whether there are any symbols, other than blanks, in the cells to the right of v' up to and including c. If there are, then $\mathcal{M}$ has not completed a computation on the input v; otherwise, it has and the output of that computation is v'.

(1.3.1) *No.* Consider the Turing machine with input alphabet $\{1\}$ and the state diagram in Figure 14. The following sequence of admissible configurations has no two terms the same:

$$(\Lambda, q_0, 1), (\mathbf{B}, q_1, \Lambda), (\mathbf{BB}, q_1, \Lambda), (\mathbf{BBB}, q_1, \Lambda), \ldots$$

(1.3.2)   The Turing machine $\mathcal{M}$ has four states. Given the input $w \equiv x_1 \ldots x_N$ in $\{0, 1\}^*$, it reads x_1 in the start state q_0, writes $\mathbf{B}$, moves R, and passes to the state

$$\begin{array}{ll} q_1 & \text{if} \quad x_1 = 0, \\ q_2 & \text{if} \quad x_1 = 1. \end{array}$$

On reading $x \in \{0, 1\}$ in the state q_i $(i = 1, 2)$, $\mathcal{M}$ writes

$$\begin{array}{ll} 0 & \text{if} \quad i = 1, \\ 1 & \text{if} \quad i = 2, \end{array}$$

moves R, and passes to the state

$$\begin{array}{ll} q_1 & \text{if} \quad x = 0, \\ q_2 & \text{if} \quad x = 1. \end{array}$$

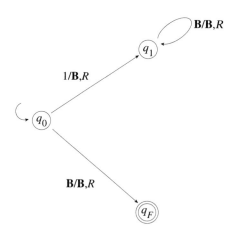

FIGURE 14. The state diagram for solution (1.3.1).

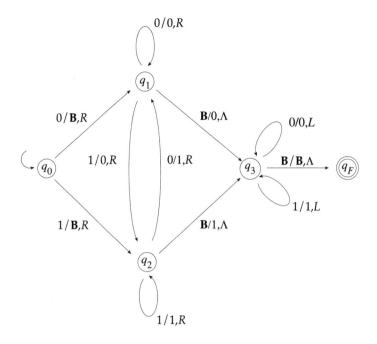

FIGURE 15. The state diagram for solution (1.3.2).

If $\mathcal{M}$ reads $\mathbf{B}$ in state q_i $(i = 1, 2)$, it writes

$$
\begin{array}{ll}
0 & \text{if} \quad i = 1, \\
1 & \text{if} \quad i = 2,
\end{array}
$$

does not move, and passes to the state q_3. If $\mathcal{M}$ reads either 0 or 1 in the state q_3, it leaves that symbol untouched, moves L, and remains in the state q_3. If $\mathcal{M}$ reads $\mathbf{B}$ in the state q_3, it leaves $\mathbf{B}$ untouched, does not move, and passes to the halt state q_F.

The state diagram for $\mathcal{M}$ is given in Figure 15.

(1.3.3) In its start state q_0, $\mathcal{M}$ reads and rewrites the leftmost symbol 0 of the word $v \in 0\mathbf{B}^*11^*$, moves R, and enters the state q_1. In that state it continues moving right, reading and rewriting $\mathbf{B}$, until it reads 1, at which stage it rewrites 1 and moves R into the state q_2.

Now suppose $\mathcal{M}$ reads 1 in the state q_2. Remaining in that state, it continues moving right, reading and rewriting 1, until it reads $\mathbf{B}$ in the cell immediately to the right of the string v. It then writes $\mathbf{B}$ and moves L into the state q_3; reads 1, writes $\mathbf{B}$, and moves L into the state q_4; reads 1, writes 1, and, remaining in the state q_4, moves L. If, at this stage, $\mathcal{M}$ reads $\mathbf{B}$, then there remain 1's to be shifted left on the tape, so $\mathcal{M}$ writes 1 and moves R into the state q_2, ready to read 1. On the other hand, if $\mathcal{M}$

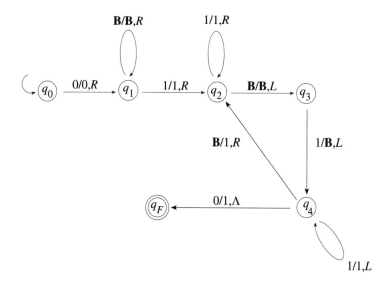

FIGURE 16. The state diagram for solution (1.3.3).

reads 0 in the state q_4, then it writes 1, does not move, and enters its halt state q_F.

The state diagram for $\mathcal{M}$ is given in Figure 16.

(1.3.4)   A Turing machine $\mathcal{M}$ with the required property behaves as follows. When started in its start state q_0 with the input word

$$w \equiv x_1 \ldots x_N,$$

reads x_1 ("memorising" it by entering an appropriate state),
writes **B**, and
moves right, one cell at a time, until a blank symbol is reached.

It then moves left, reads and memorises x_N, and writes x_1; moves left, reads and memorises x_{N-1}, and writes x_N; and so on. When the left blank is reached, it writes x_2 and passes to its halt state q_F. Figure 17 fills in the details.

Solutions for Chapter 2

(2.1.1)   The Turing machine in Figure 18 can neither leave its start state nor enter its halt state, and so computes the empty partial function $\epsilon : \mathbf{N} \to \mathbf{N}$.

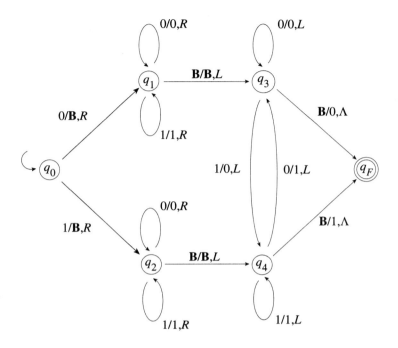

FIGURE 17. The state diagram for solution (1.3.4).

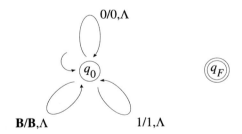

FIGURE 18. A Turing machine that computes ϵ.

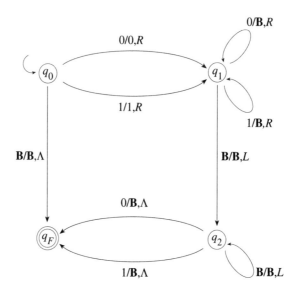

FIGURE 19. A Turing machine that computes **erase**.

(2.1.2)   See Figure 19.

(2.1.3) (i)   The Turing machine for $\wedge$ is described in Figure 20.

(2.1.3) (ii)   The Turing machine for $\vee$ is described in Figure 21.

(2.1.3) (iii)   The effect of the negation function $\neg$ on a binary string w is to convert the 1's to 0's, and the 0's to 1's. The Turing machine in Figure 22 performs this task.

(2.4.2)   We first delete the arrows representing state transitions of the form

$$\delta(q_0, y) = (q', y', D)$$

where $y \notin \{0, 1\}$. This leaves us with the state diagram in Figure 23. We could execute at this stage an encoding, etc., corresponding to the proof of Lemma (2.2). However, a smarter way to proceed is to observe that the arrow labelled $3/3, R$ and joining the state q_2 to itself makes no contribution to the computation if the input is a word over the alphabet $\{0, 1\}$. Thus, for our present purpose, we may delete that arrow from the diagram in Figure 23. If we do this, and then restrict the input and tape alphabets to $\{0, 1\}$ and $\{0, 1, 2, \mathbf{B}\}$, respectively, we are left with the same Turing machine as we dealt with in the preceding exercise. Reference to that exercise completes our solution of this one.

(2.5)   First note that the base functions are obtained from the base functions by 0 applications of composition and primitive recursion. Sup-

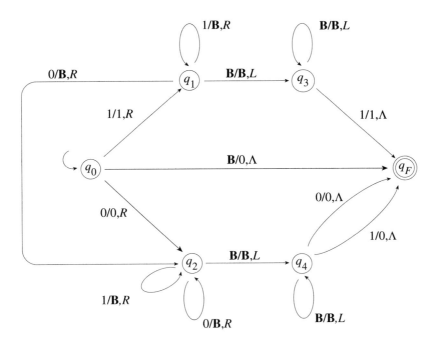

FIGURE 20. A Turing machine for ∧.

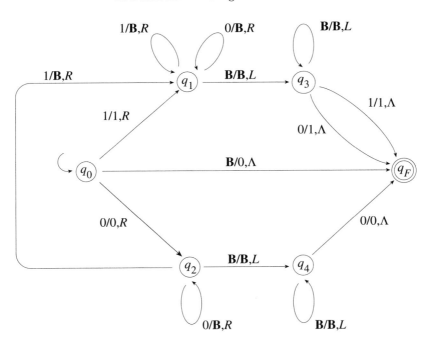

FIGURE 21. A Turing machine for ∨.

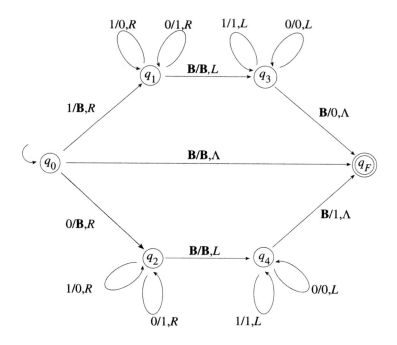

FIGURE 22. A Turing machine for ¬.

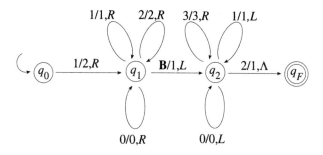

FIGURE 23. An intermediate stage in solution (2.4.2).

pose that, for some natural number k, all functions obtained from the base functions by at most k applications of the operations of composition and primitive recursion are in $\mathcal{P}$. Consider a function $f : \mathbf{N}^n \to \mathbf{N}$ obtained from the base functions by at most $k + 1$ applications of those operations. If, for example, the last of those applications in the construction of f is one of primitive recursion using functions $g : \mathbf{N}^{n-1} \to \mathbf{N}$ and $h : \mathbf{N}^{n+1} \to \mathbf{N}$, then both g and h can be obtained from the base functions by at most k applications of composition and primitive recursion, so they belong to $\mathcal{P}$, by our induction hypothesis; hence f belongs to $\mathcal{P}$ by the definition of $\mathcal{P}$. A similar argument disposes of the case where the last application of composition or primitive recursion in the construction of f is one of composition. Thus, by induction, $\mathcal{P}$ contains all functions that are constructed from the base functions by finitely many applications of composition and primitive recursion.

The reverse inclusion is an immediate consequence of the inductive definition of $\mathcal{P}$.

(2.6.1) The following recursion scheme shows that the factorial function is primitive recursive:

$$
\begin{aligned}
0! &= 1, \\
(n+1)! &= \mathbf{times} \circ (\mathbf{scsr} \circ P_1^2, P_2^2)(n, n!).
\end{aligned}
$$

(2.6.2) For each $k \in \mathbf{N}$ let f_k be the constant function $n \mapsto k$. We have the following recursion scheme for the power function:

$$
\begin{aligned}
\mathbf{power}(0, n) &= f_1(n), \\
\mathbf{power}(m + 1, n) &= \mathbf{times} \circ (P_2^3, P_3^3)(m, \mathbf{power}(m, n), n).
\end{aligned}
$$

It follows that $\mathbf{power}'$ is also primitive recursive, as

$$
\mathbf{power}' = \mathbf{power} \circ (P_2^2, P_1^2).
$$

Finally, if $m \in \mathbf{N}$ is fixed, then for each $n \in \mathbf{N}$ we have

$$
n^m = \mathbf{power} \circ (f_m, P_1^1)(n).
$$

As power, f_m, and P_1^1 are primitive recursive, so is the function $n \mapsto n^m$.

(2.6.3) First note the following recursion scheme for the function

$$
m \mapsto \mathbf{cutoff}(m, 1)
$$

on $\mathbf{N}$:

$$
\begin{aligned}
\mathbf{cutoff}(0, 1) &= 0, \\
\mathbf{cutoff}(m + 1, 1) &= P_1^2(m, \mathbf{cutoff}(m, 1)).
\end{aligned}
$$

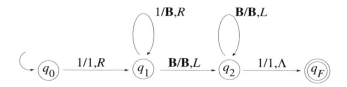

FIGURE 24. A Turing machine that computes the zero function.

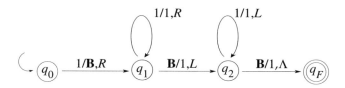

FIGURE 25. A Turing machine that computes **scsr**.

Now define an element F of $\mathcal{P}$ by the recursion scheme

$$
\begin{aligned}
F(0, n) &= P_1^1(n), \\
F(m+1, n) &= \textbf{cutoff}(P_2^3(m, F(m, n), n), 1).
\end{aligned}
$$

It is easy to show that $F(m, n) = \textbf{cutoff}(n, m)$; so the cutoff subtraction function is $F \circ (P_2^2, P_1^2)$, which certainly belongs to $\mathcal{P}$.

Since

$$|m - n| = \textbf{plus}(\textbf{cutoff}(m, n), \textbf{cutoff}(n, m))$$

and each of the functions **plus**, $(m, n) \mapsto \textbf{cutoff}(m, n)$, and $(m, n) \mapsto \textbf{cutoff}(n, m)$ is primitive recursive, so is the function $(m, n) \mapsto |m - n|$.

(2.6.4) By Exercise (2.6.2), the function $\textbf{sq} : \mathbf{N} \to \mathbf{N}$, defined by $\textbf{sq}(n) \equiv n^2$, is primitive recursive. Since, by Exercise (2.6.3), the function $(m, n) \mapsto |m - n|$ is primitive recursive, it follows that the function $f : \mathbf{N}^2 \to \mathbf{N}$ defined by $f(m, n) \equiv |m^2 - n|$ is primitive recursive. It is easy to show that $\textbf{sqrt}$ is the partial function obtained from f by minimization. Hence $\textbf{sqrt} \in \mathcal{R}$.

(2.7.1) As stated in the text, we take the natural numbers, considered as functions of zero variables, to be computable *by convention*.

The binary Turing machine in Figure 24 computes the zero function on $\mathbf{N}$.

The one in Figure 25 computes $\textbf{scsr} : \mathbf{N} \to \mathbf{N}$.

Consider the computation of $P_j^n(k_1, \ldots, k_n)$, where, for example, $1 < j < n$. The desired binary Turing machine $\mathcal{M}$ behaves as follows. Suppose the input string $k_1 0 k_2 0 ... 0 k_n$, with each $k_i \in \mathbf{N}$, is written in the left cells

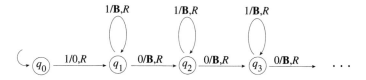

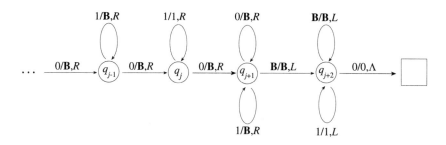

FIGURE 26. A Turing machine that computes $P_j^n(k_1, \ldots, k_n)$.

of the tape, that $\mathcal{M}$ is in its start state q_0, and that the read/write head scans the leftmost cell. $\mathcal{M}$ first writes 0 as a left end marker, and moves right, writing blanks, until it reads and deletes the $(j-1)^{\text{th}}$ instance of 0 from the original input string. It then continues moving right, leaving the content of each cell unchanged, until it reads the next instance of 0, which it deletes before moving right. It continues moving right, this time deleting all symbols until it reads $\mathbf{B}$. At that stage it moves left, leaving the content of each cell unchanged, until it reaches the 0 in the leftmost cell. It then copies the symbols of k_j onto the left of the tape, deletes all remaining nonblank symbols, and parks the read/write head.

$\mathcal{M}$ is described by the state diagram in Figure 26, in which the large square is a Turing machine module which moves the symbols of k_j to the left of the tape (cf. Exercise (1.3.3)).

(2.7.2)   Let $\mathcal{T}_i$ be a binary Turing machine that computes θ_i ($1 \leq i \leq m$), and let $\mathcal{T}$ be a binary Turing machine that computes ψ. A binary Turing machine $\mathcal{M}$ that computes the composite function $\psi \circ (\theta_1, \ldots, \theta_m)$ behaves as follows. Let $(k_1, \ldots, k_n) \in \mathbf{N}^n$. Assume that $k_1 0 \ldots 0 k_n$ is written in the left cells of the tape, that the read/write head scans the leftmost cell, and that $\mathcal{M}$ is in its start state. To begin with, $\mathcal{M}$ shifts $k_1 0 \ldots 0 k_n$ one place right, writing $\mathbf{B}$ in the leftmost cell (to act as a left end marker), copies $00 k_1 0 \ldots 0 k_n$ to the right of the tape, and places the read/write head against the cell c_1 to the right of 00. It then calls a Turing machine module that, without affecting the cells to the left of c_1, imitates the action of $\mathcal{T}_1$ on the rightmost instance of the string $k_1 0 \ldots 0 k_n$; if $\theta_1(k_1, \ldots, k_n)$ is defined, the module writes it in the cells on the right of 00 and places the read/write head against c_1. Copying $00k_10\ldots0k_n$ on the right of $\theta_1(k_1,\ldots,k_n)$ so that the tape contains the string

$$\mathbf{B}k_10\ldots0k_n00\theta_1(k_1,\ldots,k_n)00k_10\ldots0k_n,$$

$\mathcal{M}$ then places the read/write head against the cell c_2 to the right of the rightmost instance of 00 and calls a module that, without affecting the cells to the left of c_2, imitates the action of $\mathcal{T}_2$ on the rightmost instance of the string $k_10\ldots0k_n$; if $\theta_2(k_1,\ldots,k_n)$ is defined, the module writes it in the cells on the right of the rightmost instance of 00 and places the read/write head against c_2; the string on the tape at this stage is

$$\mathbf{B}k_10\ldots0k_n00\theta_1(k_1,\ldots,k_n)00\theta_2(k_1,\ldots,k_n).$$

Carrying on in this way, if $\theta_1(k_1,\ldots,k_n),\theta_2(k_1,\ldots,k_n),\ldots,\theta_m(k_1,\ldots,k_n)$ are all defined, $\mathcal{M}$ eventually arrives at a configuration in which the string on the tape is

$$\mathbf{B}k_10\ldots0k_n00\theta_1(k_1,\ldots,k_n)00\theta_2(k_1,\ldots,k_n)00\ldots00\theta_m(k_1,\ldots,k_n)$$

and the read/write head is scanning the cell to the right of the rightmost instance of 00. $\mathcal{M}$ now moves left, leaving each cell unchanged, until it reads $\mathbf{B}$ on the far left. It replaces this by 0 and moves right, writing blanks in each cell, until it reads the leftmost unit of $\theta_1(k_1,\ldots,k_n)$ (following the first instance of 00 on the tape). It then writes the string

$$\theta_1(k_1,\ldots,k_n)0\theta_2(k_1,\ldots,k_n)0\ldots0\theta_m(k_1,\ldots,k_n)$$

on the left of the tape (cf. the solution to Exercise (1.3.3)) and places the read/write head against the leftmost cell.

Finally, $\mathcal{M}$ calls a module that imitates the action of $\mathcal{T}$ on that string; so if

$$\psi\circ(\theta_1(k_1,\ldots,k_n),\theta_2(k_1,\ldots,k_n),\ldots,\theta_m(k_1,\ldots,k_n))$$

is defined, it is written on the left of the tape and $\mathcal{M}$ parks the read/write head.

(2.7.3) Let $\mathcal{T}$ be a binary Turing machine that computes ψ. A binary Turing machine $\mathcal{M}$ that computes the function φ obtained from ψ by minimization behaves as follows. Let $(k_1,\ldots,k_n)\in\mathbf{N}^n$. Assume that $k_10\ldots0k_n$ is written in the left cells of the tape, that the read/write head scans the leftmost cell, and that $\mathcal{M}$ is in its start state. To begin with, $\mathcal{M}$ shifts $k_10\ldots0k_n$ one place right, leaving $\mathbf{B}$ in the leftmost cell and placing the read/write head against that cell. It then writes 001 on the right of $k_10\ldots0k_n$, leaves the read/write head scanning the cell c on the right of 00, and enters a special state q.

Now suppose that for some $j \in \mathbf{N}$ the tape contains the string

$$\mathbf{B}k_1 0 \ldots 0 k_n 00 j$$

and that $\mathcal{M}$ is in the state q with the read/write head scanning the cell c. $\mathcal{M}$ then calls a module that, without affecting the cells to the left of c, writes $0 k_1 0 \ldots 0 k_n$ on the right of the tape and enters another special state q', with the read/write head against c. $\mathcal{M}$ now calls a module that, again without affecting the cells to the left of c, imitates the action of $\mathcal{T}$ on the string $j 0 k_1 0 \ldots 0 k_n$. If

$$(j, k_1, \ldots, k_n) \in \text{domain}(\psi),$$

this module writes $j 00 \psi(j, k_1, \ldots, k_n)$ on the right of 00 and places the read/write head against the cell c' immediately to the right of the rightmost instance of 00. $\mathcal{M}$ then calls a module that, without affecting the cells to the left of c', checks whether $\psi(j, k_1, \ldots, k_n)$ equals 0. If $\psi(j, k_1, \ldots, k_n) \neq 0$, this module writes $j+1$ in c and the cells to its right, leaves all cells further to the right blank, places the read/write head against c, and enters the state q. If $\psi(j, k_1, \ldots, k_n) = 0$, the module writes j in c and the cells to its right, leaves blanks in all cells to the right of that, and puts $\mathcal{M}$ in a special state q'' with the read/write head scanning c.

Finally, suppose that $\mathcal{M}$ is in the state q'', with the read/write head scanning the cell c, and with a string of the form j, where $j \in \mathbf{N}$, written in c and the cells to its right. $\mathcal{M}$ then calls a module that copies j onto the left of the tape, leaves all other cells blank, and halts with the read/write head on the left.

(2.9.1) $A(0, n)$ is certainly defined for all n. Suppose that $A(m, n)$ is defined for all n; then $A(m+1, 0) = A(m, 1)$ is defined. Now suppose that $A(m+1, k)$ is defined; then

$$A(m+1, k+1) = A(m, A(m+1, k))$$

is defined. Hence, by induction, $A(m+1, n)$ is defined for all n. In turn, it follows by induction that $A(m, n)$ is defined for all m and n.

(2.9.3)   We first have

$$
\begin{aligned}
A(1, 0) &= A(0, 1) &&= 2, \\
A(1, 1) &= A(0, A(1, 0)) &&= A(0, 2) &&= 3, \\
A(1, 2) &= A(0, A(1, 1)) &&= A(0, 3) &&= 4,
\end{aligned}
$$

and generally, by a simple induction argument,

$$A(1, n) = n + 2.$$

Next,

$$
\begin{aligned}
A(2,0) &= A(1,1) &&= 3, \\
A(2,1) &= A(1,A(2,0)) &&= A(1,3) &&= 5, \\
A(2,2) &= A(1,A(2,1)) &&= A(1,5) &&= 7,
\end{aligned}
$$

and, generally,

$$A(2,n) = 2n + 3.$$

Likewise,

$$
\begin{aligned}
A(3,0) &= A(2,1) &&= 5, \\
A(3,1) &= A(2,A(3,0)) &&= A(2,5) &&= 13, \\
A(3,2) &= A(2,A(3,1)) &&= A(2,13) &&= 29,
\end{aligned}
$$

and, again by induction on n,

$$A(3,n) = 2^{n+3} - 3.$$

It follows that

$$A(4,0) = A(3,1) = 13 = 2^{2^2} - 3.$$

Now suppose that, for some n,

$$A(4,n) = 2^{2^{\cdot^{\cdot^{\cdot^{2}}}}} - 3,$$

where there are $n + 3$ instances of 2 on the right hand side. Then

$$A(4,n+1) = A(3,A(4,n)) = 2^{A(4,n)+3} - 3 = 2^{2^{\cdot^{\cdot^{\cdot^{2}}}}} - 3,$$

where there are $(n+1)+3$ instances of 2 on the far right of these equations. This completes an inductive proof of the desired result.

Solutions for Chapter 3

(**3.1**)   Let S, T be recursively enumerable subsets of $\mathbf{N}$. If either S or T is empty, then it is immediate that both $S \cup T$ and $S \cap T$ are recursively enumerable; so we may assume that both S and T are nonempty. Thus there exist total computable functions s, t from $\mathbf{N}$ onto S and T, respectively. Define a total computable function f on $\mathbf{N}$ as follows: for each n, $f(2n) \equiv s(n)$ and $f(2n+1) \equiv t(n)$. Then f maps $\mathbf{N}$ onto $S \cup T$, which is therefore recursively enumerable. An effective listing of the elements of $S \cup T$ is

$$f(0), f(1), f(2), f(3), \ldots;$$

that is,

$$s(0), t(0), s(1), t(1), \ldots.$$

To construct an effective listing of the elements of $S \cap T$, follow the arrows through the diagram below, deleting all pairs $(s(i), t(j))$ with unequal components and listing the first components of the remaining pairs.

$(s(0), t(0))$ $\rightarrow$ $(s(1), t(0))$ $(s(2), t(0))$ $\rightarrow$ $(s(3), t(0))$ $\ldots$
                    $\swarrow$ $\nearrow$ $\swarrow$
$(s(0), t(1))$ $(s(1), t(1))$ $(s(2), t(1))$ $\ldots$
     $\downarrow$ $\nearrow$ $\swarrow$
$(s(0), t(2))$ $(s(1), t(2))$ $\ldots$
                    $\swarrow$
$(s(0), t(3))$ $\ldots$
     $\downarrow$
     $\vdots$

(3.4.1)   To compute $h(i, j)$, we run the Turing machine $\mathcal{M}$ on the input i and check whether it halts in at most j steps. If it does, we set $h(i, j) \equiv i$; if it does not, we set $h(i, j) \equiv a$. Intuitively, this procedure gives an algorithm for computing h. It follows from the Church-Markov-Turing thesis that h is computable.

(3.4.2)   By the definition of *recursively enumerable,* the empty subset of $\mathbf{N}$, which is the range of the empty partial function[1] on $\mathbf{N}$, is recursively enumerable, and each nonempty recursively enumerable subset of $\mathbf{N}$ is the range of a total computable function on $\mathbf{N}$. Conversely, suppose $S \subset \mathbf{N}$ is the range of a computable partial function φ from $\mathbf{N}$ to $\mathbf{N}$; we may assume that φ is not the empty partial function. By Theorem (3.3), there exists a total computable function $f : \mathbf{N} \to \mathbf{N}$ that maps $\mathbf{N}$ onto domain(φ). Since $\varphi \circ f$ is a total computable function mapping $\mathbf{N}$ onto S, it follows that S is recursively enumerable.

(3.5.1)   Let S be an infinite recursive subset of $\mathbf{N}$. Define a total function $f : \mathbf{N} \to \mathbf{N}$ inductively, as follows:

$$f(0) \equiv \text{the least } n \text{ such that } n \in S,$$

and for each $k \in \mathbf{N}$,

$$f(k + 1) \equiv \text{the least } n \text{ such that } n > f(k) \text{ and } n \in S.$$

---

[1]Recall from Exercise (2.1.1) that the empty partial function from $\mathbf{N}$ to $\mathbf{N}$ is computable.

Since S is recursive, f is computable, by the Church-Markov-Turing thesis. Also, f is strictly increasing and therefore one-one. It is clear that if $s \in S$, then $s = f(k)$ for some $k \leq s$; so f maps $\mathbf{N}$ onto S. Moreover, φ is informally computable: to compute $\varphi(n)$, we first check whether or not n belongs to S; if it does, then, by examining the values $f(0), \ldots, f(n)$, we can find $f^{-1}(n)$. By the Church-Markov-Turing thesis, φ is computable.

(3.5.2)   Let S be a recursive set. If $S = \emptyset$, then it is certainly recursively enumerable. If $S \neq \emptyset$, choose an element a of S and define a total computable function f from $\mathbf{N}$ onto S by

$$\begin{aligned} f(n) &= n \quad \text{if } n \in S, \\ &= a \quad \text{if } n \notin S. \end{aligned}$$

Then f is a recursive enumeration of S, which is therefore recursively enumerable.

(3.5.3)   Since the empty subset of $\mathbf{N}$ is recursively enumerable and since it is both the domain and the range of the (computable) empty partial function on $\mathbf{N}$, we may restrict our attention to a nonempty subset S of $\mathbf{N}$. Suppose there exists a computable partial function $\varphi : \mathbf{N} \to \mathbf{N}$ whose domain is a recursive subset of $\mathbf{N}$ and whose range is S. Fixing $a \in S$, define

$$\begin{aligned} f(n) &= \varphi(n) \quad \text{if } n \in \text{domain}(\varphi), \\ &= a \qquad \text{if } n \notin \text{domain}(\varphi). \end{aligned}$$

Then f is a total computable function from $\mathbf{N}$ onto S, which is therefore recursively enumerable. The converse is a trivial consequence of the definition of *recursively enumerable*.

(3.7.1)   First form the state transition table:

	0	**1**	**B**
0	$(1, 0, R)$	undefined	undefined
1	$(3, 1, R)$	$(1, 0, L)$	$(2, 0, R)$
2	undefined	undefined	$(2, 0, R)$

(Note that we do not need a row corresponding to the halt state 3.) Next form the string

$$\sigma \equiv 3/(1, 0, R)/\perp/\perp/(3, 1, R)/(1, 0, L)/(2, 0, R)/\perp/\perp/(2, 0, R).$$

Then encode σ :

10011	11111	10001	11111	10000	11111	11101	11111
3	/	1	/	0	/	R	/

11011	11111	11011	11111	10011	11111	10001	11111
$\perp$	/	$\perp$	/	3	/	1	/

11101	11111	10001	11111	10000	11111	11100	11111
R	/	1	/	0	/	L	/

10010	11111	10000	11111	11101	11111	11011	11111
2	/	0	/	R	/	$\perp$	/

11011	11111	10010	11111	10000	11111	11101
$\perp$	/	2	/	0	/	R

(3.7.2) (i)

10011	11111	10001	11111	10001	11111
3	/	1	/	1	/

11101	11111	11011	11111	10001
R	/	$\perp$	/	1

The string is not in range(γ), as the last symbol, 1, of its decoded form is part of an uncompleted triple.

(3.7.2) (ii)

10011	11111	10001	11111	11010	11111	11101	11111
3	/	1	/	**B**	/	R	/

10010	11111	10001	11111	11101	11111	11011	11111
2	/	1	/	R	/	$\perp$	/

10011	11111	11010	11111	11100	11111	11011	11111
3	/	**B**	/	L	/	$\perp$	/

11011	11111	10010	11111	10000	11111	11101	11111
$\perp$	/	2	/	0	/	R	/

10011	11111	11010	11111	11100	11111	11011
3	/	**B**	/	L	/	$\perp$

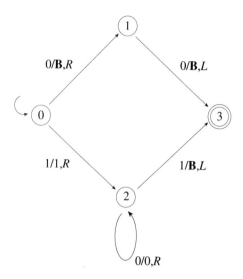

FIGURE 27. The Turing machine decoded in solution (3.7.2)(ii).

If this string is the encoded form of a normalised binary Turing machine $\mathcal{M}$, then $\mathcal{M}$ has states $0, 1, 2, 3$, with start state 0 and halt state 3. Starting from the third symbol, 1, of the decoding, we form the state transition table below. (Of course, the row corresponding to the halt state 3 does not appear in the string encoding $\mathcal{M}$.)

	0	1	B
0	$(1, \mathbf{B}, R)$	$(2, 1, R)$	undefined
1	$(3, \mathbf{B}, L)$	undefined	undefined
2	$(2, 0, R)$	$(3, \mathbf{B}, L)$	undefined
3	undefined	undefined	undefined

We now see that our string represents the normalised binary Turing machine described in Figure 27.

(3.7.2) (iii)

10011	11111	10001	11111	10000	11111	11101	11111
3	/	1	/	0	/	R	/

11011	11111	10000	11111	10001	11111	11100
$\bot$	/	0	/	1	/	L

This number is not in range(γ), as there is not sufficient information to encode a 4-by-3 transition table (corresponding to 4 states).

(3.7.3)   By Proposition (3.2), the partial function $\theta : \mathbf{N} \to \mathbf{N}$ defined by

$$
\begin{aligned}
\theta(n) &= 1 && \text{if } n \in S, \\
&= \text{undefined} && \text{if } n \notin S
\end{aligned}
$$

is computable. Hence the product $\theta\varphi_i$ is computable. But

$$
\begin{aligned}
\theta(n)\varphi_i(n) &= \varphi_i(n) && \text{if } n \in S, \\
&= \text{undefined} && \text{if } n \notin S.
\end{aligned}
$$

(3.10.1)   Define a partial function $\Psi : \mathbf{N}^3 \to \mathbf{N}$ by

$$
\Psi(n, i, j) \equiv (\varphi_i \circ \varphi_j)(n)
$$

whenever the right side exists. Then Ψ is computable: to compute $\Psi(n, i, j)$, first run $\mathcal{M}_j$ on the input n; if this computation halts, run $\mathcal{M}_i$ on the input $\varphi_j(n)$. Now choose $\nu \in \mathbf{N}$ such that $\Psi = \varphi_\nu^{(3)}$. By the s-m-n theorem, there exists a total computable function $s : \mathbf{N}^3 \to \mathbf{N}$ such that

$$
\varphi_{s(k,i,j)} = \varphi_k^{(3)}(\cdot, i, j).
$$

Defining a total computable function $g : \mathbf{N}^2 \to \mathbf{N}$ by $g(i, j) \equiv s(\nu, i, j)$, for all i and j we have

$$
\varphi_{g(i,j)} = \varphi_{s(\nu,i,j)} = \varphi_\nu^{(3)}(\cdot, i, j) = \Psi(\cdot, i, j) = \varphi_i \circ \varphi_j.
$$

(3.11.1)   Since $\psi_0, \psi_1, \ldots$ has the universal property, the partial function $\Psi : \mathbf{N}^2 \to \mathbf{N}$ defined by $\Psi(n, k) \equiv \psi_n(k)$ is computable. Using the s-m-n property of $\psi_0', \psi_1', \ldots$, we now see that there exists a total computable function $f : \mathbf{N} \to \mathbf{N}$ such that

$$
\psi'_{f(n)} = \Psi(n, \cdot) = \psi_n
$$

for each n.

(3.11.2)   The necessity of the stated condition follows from Exercise (3.11.1). To prove its sufficiency, let $\psi_0, \psi_1, \ldots$ be an enumeration of the set of all computable partial functions from $\mathbf{N}$ to $\mathbf{N}$, and suppose that there exist total computable functions $f : \mathbf{N} \to \mathbf{N}$ and $g : \mathbf{N} \to \mathbf{N}$ such that $\psi_n = \varphi_{f(n)}$ and $\varphi_n = \psi_{g(n)}$ for each n. Since the partial function $(n, k) \mapsto \varphi_{f(n)}(k)$ on $\mathbf{N}^2$ is computable (why?), $\psi_0, \psi_1, \ldots$ has the universal property. On the other hand, if $\Phi : \mathbf{N}^2 \to \mathbf{N}$ is a computable partial function, then, by Corollary (3.9), there exists a total computable function $s : \mathbf{N} \to \mathbf{N}$ such that $\varphi_{s(n)} = \Phi(n, \cdot)$ for each n. Setting $t \equiv g \circ s$, we see that t is a total computable function on $\mathbf{N}$ such that $\psi_{t(n)} = \Phi(n, \cdot)$ for each n; hence $\psi_0, \psi_1, \ldots$ has the s-m-n property.

Solutions for Chapter 4

(4.1.1) Such a Turing machine is given by the state diagram in Figure 28.

(4.1.2) Suppose there exists such a Turing machine $\mathcal{M}$, and consider its behaviour when, in its start state, it is given a nonempty input string $w \in \{0\}^*$. Since $\mathcal{M}$ halts on the input w, there is a positive integer $N > |w|$ such that, during the computation in question, $\mathcal{M}$ never reads a cell beyond the Nth from the left. Now consider what happens when the left of the tape contains w, followed by blanks in all cells up to and including the Nth, and 0 in the $(N+1)^{\text{th}}$ cell. When started on the left, $\mathcal{M}$ will mimic its behaviour on reading w; so $\mathcal{M}$ will never reach the $(N+1)^{\text{th}}$ cell, and therefore will not delete the 0 contained therein. This contradiction completes the proof.

(4.5.1) Define a total computable function $h : \mathbf{N}^2 \to \mathbf{N}$ by

$$h(n, k) \; = \; 1 \quad \text{if } \mathcal{M}_n \text{ halts in at most } k \text{ steps on the input } n,$$
$$= \; 0 \quad \text{otherwise.}$$

By following the arrows in the diagram at the top of the next page, we can produce a recursive enumeration of the range of h.

$$
\begin{array}{llll}
h(0,0) & \to \quad h(0,1) & \qquad h(0,2) & \to \quad h(0,3) \quad \dots \\
& \swarrow & \nearrow \qquad\qquad \swarrow & \\
h(1,0) & \qquad h(1,1) & \qquad h(1,2) \quad \dots & \\
\downarrow & \nearrow & \swarrow & \\
h(2,0) & \qquad h(2,1) \quad \dots & & \\
& \swarrow & & \\
h(3,0) \quad \dots & & & \\
\downarrow & & & \\
\vdots & & &
\end{array}
$$

If we go through this list and delete all terms $h(n, k)$ with the value 0, we obtain a list, say

$$h(n_0, k_0), h(n_1, k_1), h(n_2, k_2), \dots,$$

of all the values $h(n, k)$ equal to 1. Then $n_0, n_1, n_2, \dots$ is a recursive enumeration of K.

(4.5.2) Suppose $\bar{K}$ is recursively enumerable. By Theorem (3.3), $\bar{K} = \text{domain}(\varphi_\nu)$ for some ν. Then

$$
\begin{aligned}
\nu \in \bar{K} \quad &\Leftrightarrow \quad \varphi_\nu(\nu) \text{ is defined} \\
&\Leftrightarrow \quad \nu \in K \\
&\Leftrightarrow \quad \nu \notin \bar{K},
\end{aligned}
$$

a contradiction.

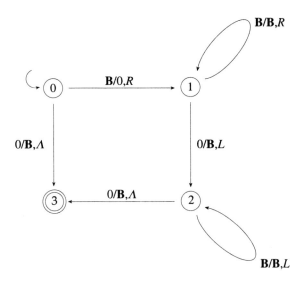

FIGURE 28. The Turing machine for solution (4.1.1).

(4.5.3)   If S is recursive, then so is $\mathbf{N}\backslash S$, as the characteristic function of the latter is $1 - \chi_S$, which is certainly computable. Then, as *recursive implies recursively enumerable* (Exercise (3.5.2)), both S and $\mathbf{N}\backslash S$ are recursively enumerable.

Conversely, suppose that both S and $\mathbf{N}\backslash S$ are recursively enumerable. If either of these sets is empty, then the other is $\mathbf{N}$ and both are certainly recursive. So we may assume that both S and $\mathbf{N}\backslash S$ are nonvoid; whence there exist total computable functions f, g on $\mathbf{N}$ such that $\mathrm{range}(f) = S$ and $\mathrm{range}(g) = \mathbf{N}\backslash S$. An informal description of an algorithm for computing χ_S goes as follows. Given $n \in \mathbf{N}$, compare it in turn with

$$f(0), g(0), f(1), g(1), \ldots$$

until either we find k with $f(k) = n$, in which case we set the output equal to 1, or else we find k such that $g(k) = n$, in which case we set the output equal to 0. Since

$$\mathrm{range}(f) \cup \mathrm{range}(g) = \mathbf{N}$$

and

$$\mathrm{range}(f) \cap \mathrm{range}(g) = \emptyset,$$

for each $n \in \mathbf{N}$ exactly one of the two alternatives for the algorithm's behaviour must occur.

Now suppose that K is recursive. Then $\bar{K}$ is recursively enumerable, by the foregoing; but this contradicts Exercise (4.5.2). Hence K is not recursive.

(4.7.1) Let $\mathcal{P}(\mathbf{N})$ denote the **power set** of $\mathbf{N}$—that is, the set of all subsets of $\mathbf{N}$. For each subset S of $\mathbf{N}$ let $F(S)$ be the number with binary expansion

$$0 \cdot \chi_S(0)\chi_S(1)\chi_S(2)\ldots.$$

Note that F maps $\mathcal{P}(\mathbf{N})$ *onto* $[0,1]$: for if $x \in [0,1]$ and $0 \cdot x_0 x_1 x_2 \ldots$ is a binary expansion of x, then $x = F(S)$, where $S \equiv \{i \in \mathbf{N} : x_i = 1\}$.

Now suppose that $\mathcal{P}(\mathbf{N})$ is countable. Then there exists a mapping g of $\mathbf{N}$ onto $\mathcal{P}(\mathbf{N})$; so $F \circ g$ is a mapping of $\mathbf{N}$ onto $[0,1]$, and therefore $[0,1]$ is countable. This contradicts Cantor's Theorem (4.6).

(4.7.2) By definition, a nonempty recursively enumerable subset S of $\mathbf{N}$ is the range of some total computable function on $\mathbf{N}$. Since the set of all computable partial functions from $\mathbf{N}$ to $\mathbf{N}$ is countable, the set of all total computable functions from $\mathbf{N}$ to $\mathbf{N}$ is also countable. Let

$$f_0, f_1, f_2, \ldots$$

be a listing[2] of all the total computable functions from $\mathbf{N}$ to $\mathbf{N}$. Then

$$\emptyset, \operatorname{range}(f_0), \operatorname{range}(f_1), \operatorname{range}(f_2), \ldots$$

is a listing of all the recursively enumerable subsets of $\mathbf{N}$; so the set of recursively enumerable subsets of $\mathbf{N}$ is countable. By Exercise (4.7.1), the set of all subsets of $\mathbf{N}$ is uncountable. Hence not all subsets of $\mathbf{N}$ are recursively enumerable—that is, there exists a subset of $\mathbf{N}$ that is not recursively enumerable.

(4.8.2) Let x be a computable real number; so there exists a total computable function $s : \mathbf{N} \to \mathbf{Q}$ such that $|x - s(k)| \leq 2^{-k}$ for each k. Given a natural number k, first compute a positive integer N such that

$$\left| e^x - \sum_{n=0}^{N} x^n/n! \right| \leq 2^{-k-1}.$$

Next compute a positive integer m such that if $|x - t| \leq 2^{-m}$, then

$$\left| \sum_{n=0}^{N} x^n/n! - \sum_{n=0}^{N} t^n/n! \right| \leq 2^{-k-1}.$$

---

[2]Note that this enumeration is not effective: see Proposition (5.1).

Then $r(k) \equiv \sum_{n=0}^{N} s(m)^n/n!$ is a rational number, and

$$
\begin{aligned}
|e^x - r(k)| &\leq \left| e^x - \sum_{n=0}^{N} x^n/n! \right| \\
&\quad + \left| \sum_{n=0}^{N} x^n/n! - \sum_{n=0}^{N} s(m)^n/n! \right| \\
&\leq 2^{-k-1} + 2^{-k-1} \\
&= 2^{-k}.
\end{aligned}
$$

We have described (informally) an algorithm for computing a total function $r : \mathbf{N} \to \mathbf{Q}$ such that $|e^x - r(k)| \leq 2^{-k}$ for each k. Hence e^x is a computable real number.

Additional Exercise: Show in detail how $r(k)$ can be constructed as a computable function of k.

(4.8.3)   Let x and y be computable real numbers, and choose total computable functions $s : \mathbf{N} \to \mathbf{Q}$ and $t : \mathbf{N} \to \mathbf{Q}$ such that $|x - s(n)| \leq 2^{-n}$ and $|y - t(n)| \leq 2^{-n}$ for each n. Define total computable functions $u, v : \mathbf{N} \to \mathbf{Q}$ by

$$
u(n) \equiv s(n+1) + t(n+1),
$$
$$
v(n) \equiv s(n+1) - t(n+1).
$$

For each n we have

$$
\begin{aligned}
|(x+y) - u(n)| &\leq |x - s(n+1)| + |y - t(n+1)| \\
&\leq 2^{-n-1} + 2^{-n-1} \\
&= 2^{-n}
\end{aligned}
$$

and similarly

$$
|(x-y) - v(n)| \leq 2^{-n}.
$$

Hence $x + y$ and $x - y$ are computable.

To handle the product xy, first compute a positive integer m such that if either $|x - z| \leq 1$ or $|y - z| \leq 1$, then $|z| \leq 2^m$. For each n we have

$$
\begin{aligned}
|xy - s(n) \cdot t(n)| &\leq |x - s(n)| \cdot |y| + |y - t(n)| \cdot |s(n)| \\
&\leq 2^{-n}2^m + 2^{-n}2^m \\
&= 2^{m-n+1}.
\end{aligned}
$$

Setting

$$
r(n) \equiv s(m+n+1) \cdot t(m+n+1),
$$

we see that $r : \mathbf{N} \to \mathbf{Q}$ is a total computable function such that $|xy - r(n)| \leq 2^{-n}$ for each n.

(4.8.4)   There exists a total computable function $s : \mathbf{N} \to \mathbf{Q}$ such that $|x - s(n)| \le 2^{-n}$ for each n. Choose a positive integer m such that $|x| > 2^{-m}$. Then $|s(n)| > 2^{-m-1}$ for all $n > m$. For such n we have

$$
\begin{aligned}
|1/x - 1/s(n)| &= |x|^{-1} |s(n)|^{-1} |x - s(n)| \\
&\le 2^m 2^{m+1} 2^{-n} \\
&= 2^{2m-n+1}.
\end{aligned}
$$

Setting

$$r(n) \equiv 1/s(2m + n + 1),$$

we see that r is a total computable function from $\mathbf{N}$ to $\mathbf{Q}$ and that

$$|1/x - r(n)| \le 2^{-n}$$

for each n.

(4.8.6)   Choosing indices i, j for φ and ψ, respectively, run $\mathcal{M}_i$ on the input n. If it completes a computation, store $\varphi(n)$ and run $\mathcal{M}_j$ on the input n. If that also completes a computation, call a Turing machine module $\mathcal{T}$ that compares the stored value of $\varphi(n)$ with $\psi(n)$ to decide whether $\varphi(n) = \psi(n)$ or $\varphi(n) \ne \psi(n)$.

(4.8.7)   Choose indices i, j for s and f, respectively. Given $n \in \mathbf{N}$, first run $\mathcal{M}_j$ on the input n, to compute and store $f(n)$. For $k = 0, 1, \dots$ use $\mathcal{M}_i$ to compute $s(k)$, and then call a Turing machine module that compares $s(k)$ with $f(n)$ (cf. the solution of Exercise (4.8.6)). If $s(k) - f(n) > 2^{-k}$, then

$$
\begin{aligned}
x - f(n) &\ge s(k) - f(n) - |x - s(k)| \\
&> 2^{-k} - 2^{-k} \\
&= 0
\end{aligned}
$$

and so $x > f(n)$; in that case we stop the computation. If $f(n) - s(k) > 2^{-k}$, then a similar argument shows that $f(n) > x$; in which case we stop the computation. On the other hand, if $|s(k) - f(n)| \le 2^{-k}$, we carry on with the computation of $s(k + 1)$. Note that we are guaranteed to find a value k such that either $s(k) - f(n) > 2^{-k}$ or $f(n) - s(k) > 2^{-k}$: for as x is irrational, there exists k such that $|x - f(n)| > 2^{-k+1}$ and therefore

$$
\begin{aligned}
|s(k) - f(n)| &\ge |x - f(n)| - |x - s(k)| \\
&> 2^{-k+1} - 2^{-k} \\
&= 2^{-k}.
\end{aligned}
$$

Thus we can be certain that our algorithm terminates.

(4.11.1)   Suppose we have a list $r_1, r_2, \ldots$ of all numbers in the closed interval $[0,1]$. Write each r_i as a decimal, and define a nonnegative real number

$$x \equiv 0 \cdot d_1 d_2 d_3 \ldots$$

by setting

$$
\begin{aligned}
d_i &= 9 \quad \text{if the } i^{\text{th}} \text{ decimal digit of } r_i \text{ is less than 5,} \\
&= 0 \quad \text{otherwise.}
\end{aligned}
$$

As x differs from r_i by 5 in the ith decimal place, it cannot be in the list $r_1, r_2, \ldots$ (see the lemma below). But this list includes *all* numbers in $[0, 1]$; so we have a contradiction from which Cantor's Theorem follows.

A diagram should help to make Cantor's argument clear:

$$
\begin{aligned}
r_0 &= \cdot\mathbf{2}\; 5\; 6\; \ldots\; 3\; \ldots \\[2mm]
r_1 &= \cdot 0\; \mathbf{6}\; 7\; \ldots\; 4\; \ldots \\[2mm]
r_3 &= \cdot 7\; 8\; \mathbf{1}\; \ldots\; 8\; \ldots \\[2mm]
&\vdots \\[1mm]
r_n &= \cdot 6\; 1\; 0\; \ldots\; \mathbf{3}\; \ldots \\[1mm]
&\vdots
\end{aligned}
$$

With the listing suggested by this diagram, since the first decimal place of r_1 is less than 5, we make the first decimal place of x equal to 9; since the second decimal place of r_2 is greater than 4, we make the second decimal place of x equal to 0; and so on. Thus

$$x = \cdot 909 \ldots 9 \ldots,$$

where the nth decimal place is 9 as r_i has n^{th} place equal to 3. Clearly, x is different from each r_i.

The lemma referred to in the above solution is the following.

Lemma.   Let $a \equiv a_0 \cdot a_1 a_2 \ldots$ and $b \equiv b_0 \cdot b_1 b_2 \ldots$ *be decimal expansions of two real numbers, and suppose that there exists n such that a_n and b_n differ by at least 2 (modulo 10). Then $a \neq b$.*

Proof.   Suppose first that $|a_0 - b_0| \geq 2$, and without loss of generality take $a_0 \geq b_0 + 2$. Then

$$a - b = a_0 - b_0 + \sum_{i=1}^{\infty}(a_i - b_i)10^{-i}$$

$$\geq 2 - \sum_{i=1}^{\infty}|a_i - b_i|\, 10^{-i}$$

$$\geq 2 - \sum_{i=1}^{\infty}9.10^{-i}$$

$$= 2 - (9/10)(1 + 1/10 + 1/10^2 + \cdots)$$

$$= 2 - \tfrac{9/10}{1-1/10} = 1;$$

whence $a \neq b$.

In general, if a_n and b_n differ by at least 2, then the integer parts of $10^n a$ and $10^n b$ differ by at least 2; so, by the case discussed above, $10^n a \neq 10^n b$, and therefore $a \neq b$.

(4.11.2)   Apply the proof, suitably adapted, of Cantor's Theorem given in the main body of the text.

(4.11.3)   **First proof.**   Since f is increasing, there exists N such that $f(n + 1) = f(n)$ for each $n \geq N$. Hence

$$x \equiv \sum_{n=0}^{\infty} 2f(n)3^{-n}$$

has a recurring ternary expansion and is therefore rational. By Exercise (4.8.1), x is computable. It follows from Lemma (4.10) that the function $2f$, and therefore f itself, is computable.

Second proof.   For each $N \in \mathbf{N}$ let f_N be the total computable function defined on $\mathbf{N}$ by

$$f_N(n) = 0 \quad \text{if } n \leq N,$$
$$= 1 \quad \text{if } n > N.$$

Clearly, f_N is computable. Since $f = f_N$ for some value of N, f is computable.

(4.14.1) (i)   A partial function $\Theta : \mathbf{R}_c \to \mathbf{N}$ is computable if there exists a computable partial function $\theta : \mathbf{N} \to \mathbf{N}$ such that if φ_i is a computable real number generator converging to a limit x in the domain of Θ, then $i \in \text{domain}(\theta)$ and $\Theta(x) = \theta(i)$.

(4.14.1) (ii)   Let $q : \mathbf{N} \to \mathbf{Q}$ be the one-one mapping of $\mathbf{Q}$ onto $\mathbf{N}$ introduced on page 52. A partial function $\Theta : \mathbf{Q} \to \mathbf{R}_c$ is computable if there exists a computable partial function $\theta : \mathbf{Q} \to \mathbf{N}$ such that if $i \in \mathbf{N}$ and $q(i) \in \text{domain}(\Theta)$, then $q(i) \in \text{domain}(\theta)$ and $\varphi_{\theta(q(i))}$ is a computable real number generator converging to $\Theta(q(i))$.

(4.14.1) (iii)   A partial function $\Theta : \mathbf{N} \times \mathbf{R}_c \to \mathbf{R}_c$ is computable if there exists a computable partial function $\theta : \mathbf{N} \times \mathbf{N} \to \mathbf{N}$ such that if $i \in \mathbf{N}$, and φ_j is a computable real number generator converging to a limit x such that $(i, x) \in \text{domain}(\Theta)$, then $(i, j) \in \text{domain}(\theta)$ and $\varphi_{\theta(i,j)}$ is a computable real number generator converging to $\Theta(i, x)$.

(4.14.2)   Define a computable partial function $\Psi : \mathbf{N}^3 \to \mathbf{N}$ as follows.

$$
\begin{aligned}
\Psi(i, j, n) &= \varphi_i(n+1) + \varphi_j(n+1) && \text{if } \varphi_i(n+1), \varphi_j(n+1) \text{ are} \\
& && \text{both defined,} \\
&= \text{undefined} && \text{otherwise.}
\end{aligned}
$$

By the s-m-n theorem, there exists a total computable function $s : \mathbf{N}^2 \to \mathbf{N}$ such that $\Psi(i, j, \cdot) = \varphi_{s(i,j)}$ for all i, j. Now consider computable real number generators φ_i, φ_j converging to computable real numbers x and y, respectively. The solution to Exercise (4.8.3) shows that $\varphi_{s(i,j)}$ is a computable real number generator converging to $x + y$. It follows that the addition function **plus** $: \mathbf{R}_c \times \mathbf{R}_c \to \mathbf{R}_c$ is computable.

A similar argument proves the computability of **minus** $: \mathbf{R}_c \times \mathbf{R}_c \to \mathbf{R}_c$. It remains to deal with **times** $: \mathbf{R}_c \times \mathbf{R}_c \to \mathbf{R}_c$. To this end, we first observe that if φ_i, φ_j are computable real number generators converging to computable real numbers x and y, respectively, then for each k,

$$
\begin{aligned}
|xy - \varphi_i(k)\varphi_j(k)| &\leq |x - \varphi_i(k)| \cdot |y| + |y - \varphi_j(k)| \cdot |\varphi_i(k)| \\
&\leq 2^{-k}(|\varphi_j(k)| + 1) + 2^{-k} |\varphi_i(k)| \\
&= 2^{-k}(1 + |\varphi_i(k)| + |\varphi_j(k)|).
\end{aligned}
$$

Define a partial function $\Psi : \mathbf{N}^3 \to \mathbf{N}$ by

$$
\Psi(i, j, n) \equiv \min k \left[2^{-k}(1 + |\varphi_i(k)| + |\varphi_j(k)|) \leq 2^{-n} \right].
$$

Ψ is computable in view of Exercises (2.6.5) and (2.7.3). Thus

$$
(i, j, n) \mapsto \varphi_i(\Psi(i, j, n)) \cdot \varphi_j(\Psi(i, j, n))
$$

is a computable partial function from $\mathbf{N}^3$ to $\mathbf{Q}$. By the s-m-n theorem, there exists a total computable function $f : \mathbf{N}^2 \to \mathbf{N}$ such that

$$\varphi_{f(i,j)} = \varphi_i(\Psi(i,j,\cdot)) \cdot \varphi_j(\Psi(i,j,\cdot))$$

for all i, j.

Now consider computable real number generators φ_i, φ_j converging to computable real numbers x and y, respectively. Since

$$
\begin{aligned}
1 + |\varphi_i(k)| + |\varphi_j(k)| &\leq 1 + (|x| + 2^{-k}) + (|y| + 2^{-k}) \\
&\leq |x| + |y| + 3,
\end{aligned}
$$

we see that

$$2^{-k}\left(1 + |\varphi_i(k)| + |\varphi_j(k)|\right) \to 0 \text{ as } k \to \infty.$$

It follows that $\varphi_{f(i,j)}$ is a total computable function on $\mathbf{N}$. Moreover, by the observation above,

$$
\begin{aligned}
\left|xy - \varphi_{f(i,j)}(n)\right| &= |xy - \varphi_i(\Psi(i,j,n)) \cdot \varphi_j(\Psi(i,j,n))| \\
&\leq 2^{-\Psi(i,j,n)}\left(1 + |\varphi_i(\Psi(i,j,n))| + |\varphi_j(\Psi(i,j,n))|\right) \\
&\leq 2^{-n}
\end{aligned}
$$

for each n. Hence $\varphi_{f(i,j)}$ is a computable real number generator converging to xy.

(4.14.3)   Let $\mathcal{M}$ be a normalised binary Turing machine that computes ψ. To compute $\Psi(m, n)$, first run $\mathcal{M}$ on the input m. If $\mathcal{M}$ completes the computation, run $\mathcal{M}_m$ on the input $2\psi(m) + n - 2$. If that computation is completed, run a Turing machine module that checks whether or not $\varphi_m(2\psi(m) + n - 2) = 0$; finally, if $\varphi_m(2\psi(m) + n - 2) \neq 0$, run a Turing machine module that computes $1/\varphi_m(2\psi(m) + n - 2)$.

(4.14.4)   There exist computable partial functions $\theta : \mathbf{N} \to \mathbf{N}$ and $\psi : \mathbf{N} \to \mathbf{N}$ such that

(i)   if φ_i is a computable real number generator converging to a computable real number $x \in \text{domain}(\Theta)$, then $i \in \text{domain}(\theta)$ and $\varphi_{\theta(i)}$ is a computable real number generator converging to $\Theta(x)$;

(ii)   if $i \in \text{domain}(\Psi)$, then $i \in \text{domain}(\psi)$ and $\varphi_{\psi(i)}$ is a computable real number generator converging to $\Psi(i)$.

If $i \in \text{domain}(\Theta \circ \Psi)$, then $i \in \text{domain}(\Psi)$ and $\Psi(i) \in \text{domain}(\Theta)$. So $i \in \text{domain}(\psi)$ and $\varphi_{\psi(i)}$ is a computable real number generator converging to the computable real number $\Psi(i)$; whence $\psi(i) \in \text{domain}(\theta)$ and $\varphi_{\theta(\psi(i))}$ is a computable real number generator converging to the computable real number $\Theta(\Psi(i))$. Thus $\Theta \circ \Psi$ is a computable partial function from $\mathbf{N}$ to $\mathbf{R}_c$.

On the other hand, there exists a computable partial function $\psi' : \mathbf{N} \to \mathbf{N}$ such that if φ_i is a computable real number generator converging to a computable real number $x \in \text{domain}(\Psi')$, then $i \in \text{domain}(\psi')$ and $\varphi_{\psi'(i)}$ is a computable real number generator converging to $\Psi'(x)$. Let φ_i be a computable real number generator converging to $x \in \text{domain}(\Theta \circ \Psi')$. Then $i \in \text{domain}(\psi')$, $\varphi_{\psi'(i)}$ is a computable real number generator converging to $\Psi'(x)$, and $\Psi'(x) \in \text{domain}(\Theta)$. Hence $\psi'(i) \in \text{domain}(\theta)$ and $\varphi_{\theta(\psi'(i))}$ is a computable real number generator converging to $\Theta(\Psi'(i))$. Thus $\Theta \circ \Psi'$ is a computable partial function from $\mathbf{R}_c$ to $\mathbf{R}_c$.

(4.14.5)   First note that (cf. Exercise (4.14.1)) a partial function $\Theta : \mathbf{R}_c \times \mathbf{R}_c \to \mathbf{N}$ is **computable** if there exists a computable partial function $\theta : \mathbf{N}^2 \to \mathbf{N}$ with the following property: if φ_i, φ_j are computable real number generators converging to computable real numbers x and y, respectively, such that $(x, y) \in \text{domain}(\Theta)$, then $(i, j) \in \text{domain}(\theta)$ and $\theta(i, j) = \Theta(x, y)$.

Now let $\Theta : \mathbf{R}_c \times \mathbf{R}_c \to \mathbf{N}$ be a computable partial function, fix $a \in \mathbf{R}_c$, and let Ψ be the partial function $x \mapsto \Theta(x, a)$ on $\mathbf{R}_c$. Choose the computable partial function $\theta : \mathbf{N}^2 \to \mathbf{N}$ as above, let ν be an index of a computable real number generator that converges to a, and define a computable partial function $\varphi : \mathbf{N} \to \mathbf{N}$ by $\varphi \equiv \theta(\cdot, \nu)$. If φ_i is a computable real number generator converging to a computable real number $x \in \text{domain}(\Psi)$, then $i \in \text{domain}(\varphi)$ and

$$\varphi(i) = \theta(i, \nu) = \Theta(x, a).$$

Hence θ is computable.

(4.14.6)   To begin with, suppose that Θ is computable; so there exists a computable partial function $\theta : \mathbf{N} \to \mathbf{N}$ such that if $i \in \text{domain}(\Theta)$, then $i \in \text{domain}(\theta)$ and $\varphi_{\theta(i)}$ is a computable real number generator converging to $\Theta(i)$. Define a computable partial function $\gamma : \mathbf{N} \to \mathbf{N}$ by

$$\gamma(j) \equiv \min k \left[|\varphi_j(2) - k| < 1/2 \right].$$

Define also a computable partial function $\Psi : \mathbf{N}^2 \to \mathbf{N}$ by

$$\Psi(j, n) \equiv \varphi_{\theta(\gamma(j))}(n).$$

Choose a total computable function $s : \mathbf{N} \to \mathbf{N}$ such that $\Psi(j, \cdot) = \varphi_{s(j)}$ for each j. Let φ_j be a computable real number generator converging to $i \in \mathbf{N}$. Then as $|i - \varphi_j(2)| < 1/2$, we see that $\gamma(j) = i$. Thus

$$\varphi_{s(j)} = \varphi_{\theta(\gamma(j))} = \varphi_{\theta(i)},$$

which is a computable real number generator converging to $\Theta^*(i)$. Hence Θ^* is computable.

Conversely, suppose that Θ^* is computable, and choose a computable partial function $\theta^* : \mathbf{N} \to \mathbf{N}$ such that if φ_j is a computable real number generator converging to $i \in \mathrm{domain}(\Theta^*)$, then $j \in \mathrm{domain}(\theta^*)$ and $\varphi_{\theta^*(j)}$ is a computable real number generator converging to $\Theta^*(i)$. Define a total computable function $F : \mathbf{N}^2 \to \mathbf{N}$ by

$$F(i, n) \equiv i \quad (i, n \in \mathbf{N}).$$

Applying the s-m-n theorem, construct a total computable function $t : \mathbf{N} \to \mathbf{N}$ such that $F(i, \cdot) = \varphi_{t(i)}$ for each i. Let $\theta \equiv \theta^* \circ t$; then θ is a computable partial function from $\mathbf{N}$ to $\mathbf{N}$. If $i \in \mathrm{domain}(\Theta)$, then $\varphi_{t(i)}$ is, trivially, a computable real number generator converging to i, so $t(i) \in \mathrm{domain}(\theta^*)$ and $\varphi_{\theta(i)}$ is a computable real number generator converging to $\Theta^*(i) = \Theta(i)$. Hence Θ is computable.

(4.17.1)   It is clear from the definition of the mapping q on page 52 that there exist total computable functions $f : \mathbf{N} \to \mathbf{N}$, $g : \mathbf{N} \to \mathbf{N}$, and $s : \mathbf{N} \to \{0, 1\}$ such that for each i,

$$q(i) = (-1)^{s(i)} f(i)/g(i).$$

Given $i \in \mathbf{N}$, and working relative to the number base d, we now divide $g(i)$ into $f(i)$ by the standard long division algorithm from elementary school to produce successively the digits of a d-ary expansion of $|q(i)|$. Thus

$$\Phi(i, n) \quad \equiv \quad \text{the } (n+1)^{\text{th}} \text{ digit produced by the long division of } f(i) \text{ by } g(i)$$

defines a total computable function $\Phi : \mathbf{N}^2 \to \mathbf{N}$. Using the s-m-n theorem, we now obtain a total computable function $r : \mathbf{N} \to \mathbf{N}$ such that

$$q(i) = (-1)^{s(i)} \sum_{n=0}^{\infty} \varphi_{r(i)}(n) d^{-n}$$

for each i.

(4.17.2)   We begin with the following observation from Exercise (4.8.7): *If φ_i is a computable real number generator converging to an irrational computable real number x, and $f : \mathbf{N} \to \mathbf{Q}$ is a total computable function, then*

(i)   *for each $n \in \mathbf{N}$ there exists k such that $|\varphi_i(k) - f(n)| > 2^{-k}$;*
(ii)   *if $\varphi_i(k) - f(n) > 2^{-k}$, then $x > f(n)$;*
(iii)   *if $f(n) - \varphi_i(k) > 2^{-k}$, then $x < f(n)$.*

The solution of the present exercise is modelled on the proof of Proposition (4.16). We define simultaneously computable partial functions α :

$\mathbf{N}^3 \to \mathbf{N}$ and $\Psi : \mathbf{N}^2 \to \mathbf{N}$ as follows. First set

$$\alpha(0, i, n) \equiv \min m \left[|\varphi_i(m) - n| > 2^{-m} \right],$$
$$\Psi(i, 0) \equiv \min n \left[n - \varphi_i(\alpha(0, i, n)) > 2^{-\alpha(0,i,n)} \right] - 1.$$

Having defined $\Psi(i, j)$ $(0 \le j \le k)$, if $t \in \{0, \dots, d-1\}$ set

$$\alpha(k+1, i, t) \equiv \min m \left[\left| \varphi_i(m) - \sum_{j=0}^{k} \Psi(i,j)d^{-j} - (t+1)d^{-k-1} \right| > 2^{-m} \right];$$

otherwise set

$$\alpha(k+1, i, t) \equiv \text{undefined}.$$

Define also

$$\Psi(i, k+1)$$
$$\equiv \min t \left[t \in \{0, \dots, d-1\} \text{ and } \sum_{j=0}^{k} \Psi(i,j)d^{-j} + (t+1)d^{-k-1} \right.$$
$$\left. - \varphi_i(\alpha(k+1, i, t)) > 2^{-\alpha(k+1,i,t)} \right] - 1.$$

Using the s-m-n theorem, choose a total computable function $s : \mathbf{N} \to \mathbf{N}$ such that $\varphi_{s(i)} = \Psi(i, \cdot)$ for each i. Now consider $i \in \mathbf{N}$ such that φ_i is a computable real number generator converging to a positive irrational number $x \in \mathbf{R}_c$. Taking $f(n) \equiv n$ in the observation at the start of this solution, we see that $\alpha(0, i, \cdot)$ is a total function on $\mathbf{N}$ and that $\Psi(i, 0)$ is the unique natural number N such that $N < x < N+1$. Now suppose that the values $\Psi(i, j)$ $(0 \le j \le k)$ have been computed so that

$$\Psi(i, j) \in \{0, 1, \dots, d-1\} \quad (1 \le j \le k).$$

Taking

$$f(t) = \sum_{j=0}^{k} \Psi(i,j)d^{-j} + (t+1)d^{-k-1} \quad \text{if } 0 \le t \le d-1,$$
$$= -1 \quad \text{if } t \ge d,$$

and applying the observation at the start of this solution, we see that $\alpha(k+1, i, t)$ is defined for each $t \in \{0, \dots, d-1\}$ and that $\Psi(i, k+1)$ is the unique $t \in \{0, \dots, d-1\}$ such that

$$\sum_{j=0}^{k} \Psi(i,j)d^{-j} + td^{-k-1} < x < \sum_{j=0}^{k} \Psi(i,j)d^{-j} + (t+1)d^{-k-1}.$$

It follows that

$$x = \sum_{n=0}^{\infty} \Psi(i, n)d^{-n} = \sum_{n=0}^{\infty} \varphi_{s(i)}(n)d^{-n}.$$

(4.23.1)   Define a total computable function $H : \mathbf{N}^2 \to \{0, 1\}$ by

$$
\begin{aligned}
H(m, n) &= 1 \quad \text{if } \mathcal{M}_m \text{ computes } \varphi_m(m) \text{ in at most } n \text{ steps,} \\
&= 0 \quad \text{otherwise.}
\end{aligned}
$$

By Corollary (3.9), there exists a total computable function $s : \mathbf{N} \to \mathbf{N}$ such that $\varphi_{s(m)} = H(m, \cdot)$ for each $m \in \mathbf{N}$. Suppose there exists a computable partial function θ with the stated properties. For each i, $\varphi_{s(i)}$ is a total computable function from $\mathbf{N}$ into $\{0, 1\}$, so $s(i) \in \mathrm{domain}(\theta)$. If $\theta(s(i)) = 0$, then $\varphi_{s(i)}(n) = 0$ for all n, so $\varphi_i(i)$ is undefined. If $\theta(s(i)) = 1$, then there exists n such that $\varphi_{s(i)}(n) = 1$, so $\varphi_i(i)$ is defined. Thus K is recursive, which contradicts Corollary (4.3).

Comment: It is easily seen that there is no computable partial function $\theta : \mathbf{N} \to \mathbf{N}$ such that if $\varphi_i : \mathbf{N} \to \{0, 1\}$ is an *increasing* total function, then $i \in \mathrm{domain}(\theta)$ and

$$
\begin{aligned}
\theta(i) = 0 &\quad \Rightarrow \quad \varphi_i(n) = 0 \text{ for all } n, \\
\theta(i) = 1 &\quad \Rightarrow \quad \text{there exists } n \text{ such that } \varphi_i(n) = 1.
\end{aligned}
$$

(4.23.2)   Suppose there exists a computable partial function $\theta : \mathbf{N} \to \mathbf{N}$ with the stated properties. Given i such that φ_i is a total function from $\mathbf{N}$ into $\{0, 1\}$, define

$$
x \equiv -\sum_{n=0}^{\infty} 2^{-n} \varphi_i(n).
$$

Then (cf. Exercise (4.8.5)) x is a computable real number, and

$$
s(k) \equiv -\sum_{n=0}^{k} 2^{-n} \varphi_i(n) \quad (k \in \mathbf{N})
$$

defines a computable real number generator s converging to x. Choose m such that $s = q \circ \varphi_m$. If $\theta(m) = 0$, then $x < 0$, so there exists n such that $\varphi_i(n) = 1$. If $\theta(m) = 1$, then $x \geq 0$, so $\varphi_i(n) = 0$ for all n. It follows that the value of $\theta(m)$ is independent of the index m of s and that

$$
\psi(i) \equiv 1 - \theta(m)
$$

defines a computable partial function $\psi : \mathbf{N} \to \{0, 1\}$ such that if φ_i is a total function from $\mathbf{N}$ into $\{0, 1\}$, then $i \in \mathrm{domain}(\psi)$,

$$
\begin{aligned}
\psi(m) = 0 &\quad \Rightarrow \quad \varphi_i(n) = 0 \text{ for all } n, \text{ and} \\
\psi(m) = 1 &\quad \Rightarrow \quad \text{there exists } n \text{ such that } \varphi_i(n) = 1.
\end{aligned}
$$

This contradicts Exercise (4.23.1).

(4.23.3)   Suppose such a total computable function f exists. Thus there exists a computable partial function $\psi : \mathbf{N} \to \{0, 1\}$ such that if

φ_k is a computable real number generator converging to a computable real number x, then $k \in \text{domain}(\psi)$ and $\psi(k) = f(x)$. Define a computable partial function $\Psi : \mathbf{N}^2 \to \mathbf{N}$ by

$$\Psi(i, k) \equiv \sum_{n=0}^{k} 2^{-n} \varphi_i(n).$$

Using the s-m-n theorem, construct a total computable function $s : \mathbf{N} \to \mathbf{N}$ such that $\varphi_{s(i)} = \Psi(i, \cdot)$ for each i. Then $\theta \equiv \psi \circ s$ is a computable partial function from $\mathbf{N}$ to $\{0, 1\}$. If $\varphi_i : \mathbf{N} \to \{0, 1\}$ is total, then $\varphi_{s(i)}$ is a computable real number generator converging to the computable real number

$$x \equiv \sum_{n=0}^{\infty} 2^{-n} \varphi_i(n),$$

so $s(i) \in \text{domain}(\psi)$, $i \in \text{domain}(\theta)$, and $\theta(i) = \psi(s(i)) = f(x)$. If $\theta(i) = 0$, then $x = 0$, so $\varphi_i(n) = 0$ for all n; if $\theta(i) \neq 0$, then $x \neq 0$, so $\varphi_i(n) = 1$ for some n. This contradicts Exercise (4.23.1).

There is no algorithm which, applied to any rational number x, will decide whether $x = 0$ or $x \neq 0$. To see this, suppose there is such an algorithm. Then there exists a total computable function $g : \mathbf{Q} \to \{0, 1\}$ such that

$$g(x) = 0 \quad \Rightarrow \quad x = 0,$$
$$g(x) = 1 \quad \Rightarrow \quad x \neq 0.$$

If φ_i is an increasing total computable function from $\mathbf{N}$ to $\{0, 1\}$, then

$$x \equiv \sum_{n=0}^{\infty} 2^{-n} \varphi_i(n)$$

has a recurring binary expansion and so is rational. By considering $g(x)$, we can decide whether $\varphi_i(n) = 0$ for all n or there exists N such that $\varphi_i(N) = 1$. This contradicts the comment at the end of the solution of Exercise (4.23.1). (The details are left to you.)

(4.23.5)   Suppose such a function f exists. Then there exists a computable partial function $\psi : \mathbf{N} \to \{0, 1\}$ such that if φ_m is a computable real number generator converging to the computable real number x, then $m \in \text{domain}(\psi)$ and $\psi(m) = f(x)$. Choose a strictly increasing sequence $(\nu_k)_{k=0}^{\infty}$ such that

$$\left| e - \sum_{n=0}^{\nu_k} 1/n! \right| \leq 2^{-k} \quad (k \in \mathbf{N}).$$

Define a computable partial function $\Phi : \mathbf{N}^2 \to \mathbf{N}$ by

$$\Phi(i,k) \equiv \sum_{n=0}^{\nu_k} \varphi_i(n)/n!$$

By the s-m-n theorem, there exists a total computable function $s : \mathbf{N} \to \mathbf{N}$ such that $\varphi_{s(i)} = \Phi(i,\cdot)$ for each i. Let $\theta \equiv \psi \circ s$ (a computable partial function from $\mathbf{N}$ to $\{0,1\}$). Given i such that $\varphi_i : \mathbf{N} \to \{0,1\}$ is total and increasing, we see that $\varphi_{s(i)}$ is a computable real number generator converging to the computable real number

$$x \equiv \sum_{n=0}^{\infty} \varphi_i(n)/n!$$

Note that x is irrational if and only if there exists N such that $\varphi_i(N) = 0$ and $\varphi_i(N+1) = 1$. For if there is no such N, then $\varphi_i(n) = 0$ for all n, and therefore $x = 0$; on the other hand, if such N exists, then

$$x = e - \sum_{n=0}^{N}(1/n!),$$

which, being the difference of an irrational number and a rational number, is irrational. Since $s(i) \in \mathrm{domain}(\psi)$, $i \in \mathrm{domain}(\theta)$. If $\theta(i) = 0$, then $f(x) = \psi(s(i)) = 0$; so x is rational, and therefore $\varphi_i(n) = 0$ for all n. If $\theta(i) = 1$, then $f(x) \neq 0$; so x is irrational, and therefore there exists n such that $\varphi_i(n) = 1$. This contradicts the comment at the end of the solution of (4.23.1).

(4.24.1)   Let $(x_n)_{n=0}^{\infty}$ be a computable sequence of computable real numbers that converges effectively to a real number x. So there exist total computable functions $f : \mathbf{N} \to \mathbf{N}$ and $h : \mathbf{N} \to \mathbf{N}$ such that for each n,

(i) $\varphi_{f(n)}$ is a computable real number generator converging to x_n, and

(ii)   if $k \geq h(n)$, then $|x - x_k| \leq 2^{-n}$.

Define a total computable function $g : \mathbf{N} \to \mathbf{Q}$ by

$$g(n) \equiv \varphi_{f \circ h(n+1)}(n+1).$$

Then

$$
\begin{aligned}
|x - g(n)| &\leq \left|x - x_{h(n+1)}\right| + \left|x_{h(n+1)} - \varphi_{f \circ h(n+1)}(n+1)\right| \\
&\leq 2^{-n-1} + 2^{-n-1} \\
&= 2^{-n}.
\end{aligned}
$$

Hence x is a computable real number.

(4.24.2)   Referring to Exercise (4.5.1), choose a total computable function f from $\mathbf{N}$ onto K. Define

$$a_n \equiv \sum_{m=0}^{n} 2^{-f(m)-1} \quad (n = 0, 1, 2, \ldots).$$

Then each a_n is rational and therefore computable, and $a_0 < a_1 < \cdots$. Suppose that (a_n) converges effectively to a real number a, and let $g : \mathbf{N} \to \mathbf{N}$ be a total computable function such that $|a - a_n| \le 2^{-k}$ whenever $n \ge g(k)$. If $n > g(i+2)$, then

$$
\begin{aligned}
2^{-f(n)-1} &\le a_n - a_{g(i+2)} \\
&\le |a_n - a| + |a - a_{g(i+2)}| \\
&\le 2^{-i-2} + 2^{-i-2} \\
&< 2^{-i-1}
\end{aligned}
$$

so $f(n) > i$. It follows that $i \in K$ if and only if there exists $n \le g(i+2)$ such that $f(n) = i$. Comparing i with

$$f(1), \ldots, f(g(i+2))$$

in turn, we can decide whether $i \in K$. Hence K is recursive. This contradiction completes the proof that (a_n) does not converge effectively.

Comments:

- A fundamental theorem of classical analysis is the **monotone sequence principle**: *An increasing sequence of real numbers that is bounded above converges to its least upper bound.* Specker's Theorem shows that the natural recursive analogue of that principle fails to hold.

- Proofs of the following strong version of Specker's Theorem can be found as Theorem (5.4) of [1] and Theorem (3.1) of [8]: *There exist a strictly increasing computable sequence (a_n) of rational numbers in $[0,1]$, and total computable functions $F : \mathbf{R}_c \times \mathbf{N} \to \mathbf{N}$ and $h : \mathbf{N} \to \mathbf{N}$, such that $|x - a_n| \ge 2^{-h(m)}$ whenever $x \in \mathbf{R}_c$ and $n \ge F(x, m)$.*

(4.24.3)   Choose a computable partial function $\theta : \mathbf{N}^2 \to \mathbf{N}$ such that if φ_i is a computable real number generator converging to $x \in \mathbf{R}_c$, then $(n, i) \in \text{domain}(\theta)$ and $\varphi_{\theta(n,i)}$ is a computable real number generator converging to $f_n(x)$. Define a computable partial function $\Psi : \mathbf{N}^3 \to \mathbf{Q}$ by

$$\Psi(n, i, m) = \sum_{k=0}^{n} \varphi_{\theta(k,i)}(m + k + 1).$$

Using the s-m-n theorem, construct a total computable function $g : \mathbf{N}^2 \to \mathbf{N}$ such that $\varphi_{g(n,i)} = \Psi(n, i, \cdot)$ for all n and i. Consider a computable real number generator φ_i converging to $x \in \mathbf{R}_c$. For each $n, \varphi_{g(n,i)}$ is a total computable function from $\mathbf{N}$ to $\mathbf{N}$; so for all m,

$$
\begin{aligned}
\left| s_n(x) - \varphi_{g(n,i)}(m) \right| &= \left| \sum_{k=0}^{n} f_k(x) - \sum_{k=0}^{n} \varphi_{\theta(k,i)}(m+k+1) \right| \\
&\leq \sum_{k=0}^{n} \left| f_k(x) - \varphi_{\theta(k,i)}(m+k+1) \right| \\
&\leq \sum_{k=0}^{n} 2^{-m-k-1} \\
&< 2^{-m}.
\end{aligned}
$$

Thus $\varphi_{g(n,i)}$ is a computable real number generator converging to $s_n(x)$. It follows that (s_n) is a computable sequence of total computable functions from $\mathbf{R}_c$ to $\mathbf{R}_c$.

(4.24.4)   Assume that (f_n) is a computable sequence of total computable functions from $\mathbf{R}_c$ to $\mathbf{R}_c$. So there exists a total computable function $\theta : \mathbf{N}^2 \to \mathbf{N}$ such that if φ_i is a computable real number generator converging to $x \in \mathbf{R}_c$, then $(n, i) \in \mathrm{domain}(\theta)$ and $\varphi_{\theta(n,i)}$ is a computable real number generator converging to $f_n(x)$. Define a computable partial function $\Psi : \mathbf{N}^2 \to \mathbf{Q}$ by

$$
\Psi(i, n) \equiv \varphi_{\theta(h(n+1),i)}(n+1),
$$

and use the s-m-n theorem to construct a total computable function $s : \mathbf{N} \to \mathbf{N}$ such that $\varphi_{s(i)} = \Psi(i, \cdot)$ for each i. Let x be a computable real number, and φ_i a computable real number generator converging to x. Then for each n we have

$$
\begin{aligned}
\left| \varphi_{s(i)}(n) - f(x) \right| &\leq \left| \varphi_{\theta(h(n+1),i)}(n+1) - f_{h(n+1)}(x) \right| \\
&\quad + \left| f_{h(n+1)}(x) - f(x) \right| \\
&\leq 2^{-n-1} + 2 \\
&= 2^{-n}.
\end{aligned}
$$

Thus $\varphi_{s(i)}$ is a computable real number generator converging to $f(x)$. It follows that f is a total computable function from $\mathbf{R}_c$ to $\mathbf{R}_c$.

(4.24.5)   There exists a total computable function $s : \mathbf{N} \to \mathbf{N}$ such that for each k, $\varphi_{s(k)}$ is a computable real number generator converging to x_k. Also, there exists a computable partial function $\theta : \mathbf{N}^2 \to \mathbf{N}$ such that if φ_i is a computable real number generator converging to $x \in \mathbf{R}_c$, then for each

n, $(n,i) \in \mathrm{domain}(\theta)$ and $\varphi_{\theta(n,i)}$ is a computable real number generator converging to $f_n(x)$. Now define a total computable function $T : \mathbf{N}^3 \to \mathbf{N}$ by

$$T(n,k,i) \equiv \varphi_{\theta(n,s(k))}(i),$$

and set

$$r_{n,k} \equiv T(n,k,k) \quad (n,k \in \mathbf{N}).$$

Then $(r_{n,k})_{n,k=0}^{\infty}$ is a computable double sequence of rational numbers, and for all $n,k \in \mathbf{N}$,

$$\left| f_n(x_k) - r_{n,k} \right| = \left| f_n(x_k) - \varphi_{\theta(n,s(k))}(k) \right| \leq 2^{-k}.$$

(**4.29.2**) Since f_n maps $\mathbf{Q}$ to $\mathbf{Q}$,

$$\Phi(n,i,k) \equiv f_n \circ \varphi_i \circ h(n,k)$$

defines a computable partial function $\Phi : \mathbf{N}^3 \to \mathbf{Q}$. Choose a total computable function $s : \mathbf{N}^2 \to \mathbf{N}$ such that $\varphi_{s(n,i)} = \Phi(n,i,\cdot)$ for all n,i. Consider a computable real number generator φ_i converging to $x \in \mathbf{R}_c$. For all n and k, since

$$|x - \varphi_i \circ h(n,k)| \leq 2^{-h(n,k)},$$

we have

$$\left| f_n(x) - \varphi_{s(n,i)}(k) \right| \leq 2^{-k}.$$

Hence $\varphi_{s(n,i)}$ is a computable real number generator converging to $f_n(x)$. This shows both that f_n maps $\mathbf{R}_c$ into $\mathbf{R}_c$ and that $(n,x) \mapsto f_n(x)$ is a computable partial function from $\mathbf{N} \times \mathbf{R}_c$ into $\mathbf{R}_c$.

(**4.29.3**)   We may assume that h takes only positive integer values. For all $m,n \in \mathbf{N}$ let

$$x_{m,k} \equiv -2^{m-1} + \frac{k}{2^{h(n,1)}} 2^m \quad (0 \leq k \leq 2^{h(n,1)})$$

and

$$b(m,n) \equiv 1 + \max\{|f_n(x_{m,k})| : 0 \leq k \leq 2^{h(n,1)}\}.$$

Then b is a total computable function from $\mathbf{N}^2$ to $\mathbf{N}$. (*Additional exercise*: provide a detailed proof that b is computable.) Given $m,n \in \mathbf{N}$ and a real number x with $-2^{m-1} \leq x \leq 2^{m-1}$, choose k such that $0 \leq k \leq 2^{h(n,1)}$ and $|x - x_{m,k}| \leq 2^{-h(n,1)}$. Then

$$|f_n(x)| \leq |f_n(x) - f_n(x_{m,k})| + |f_n(x_{m,k})| \leq 1 + |f_n(x_{m,k})| \leq b(m,n).$$

(4.29.4)   It readily follows from the definition of s_n that s_n maps $\mathbf{Q}$ into $\mathbf{Q}$. Noting that

$$|s_n(x) - s_n(y)| \le |x - y| \quad (n \in \mathbf{N}, \ x \in \mathbf{R}),$$

we see immediately from Exercise (4.29.2) that each s_n maps $\mathbf{R}_c$ into $\mathbf{R}_c$, and that $(s_n)_{n=0}^{\infty}$ is a computable sequence of total computable functions from $\mathbf{R}_c$ to $\mathbf{R}_c$. Since f is effectively uniformly continuous, there exists a total computable function $g : \mathbf{N} \to \mathbf{N}$ such that if $k \in \mathbf{N}$, if $x, y \in \mathbf{R}$, and if $|x - y| \le 2^{-g(k)}$, then $|f(x) - f(y)| \le 2^{-k}$. Also, applying Exercise (4.29.3) with $f_n \equiv f$ for each n, we can construct a total computable function $b : \mathbf{N} \to \mathbf{N}\backslash\{0\}$ such that

$$|f(x)| \le b(n) \quad (n \in \mathbf{N}, \ x \in [-n, n]).$$

Define a total computable function $h : \mathbf{N}^2 \to \mathbf{N}$ by

$$h(n, k) \equiv \min m \left[2^{-m} \le \min\{2^{-g(k+1)}, 2^{-k-1}b(n+3)^{-1}\} \right].$$

Consider $n, k \in \mathbf{N}$ and real numbers x, y such that $|x - y| \le 2^{-h(n,k)}$. We have

$$
\begin{aligned}
|(fs_n)(x) - (fs_n)(y)| &\le |f(x) - f(y)| \cdot |s_n(x)| \\
&\quad + |s_n(x) - s_n(y)| \cdot |f(y)| \\
&\le 2^{-k-1} + |x - y| \cdot |f(y)|.
\end{aligned}
$$

If $x \in [-n-2, n+2]$, then $y \in [-n-3, n+3]$, so that

$$|(fs_n)(x) - (fs_n)(y)| \le 2^{-k-1} + 2^{-h(n,k)}b(n+3) \le 2^{-k}.$$

On the other hand, if $x \notin [-n-2, n+2]$, then $|x| > n+1$ and $|y| > n+1$, so $s_n(x) = 0 = s_n(y)$ and therefore

$$|(fs_n)(x) - (fs_n)(y)| \le 2^{-k-1} < 2^{-k}.$$

(4.29.6)   For each $x \in \mathbf{R}$ the series $\sum_{n=0}^{\infty} 2^{-n}t_n(x)$ converges by comparison with $\sum_{n=0}^{\infty} 2^{-n}$; moreover, since

$$\left| t(x) - \sum_{n=0}^{m} 2^{-n}t_n(x) \right| \le \sum_{n=m+1}^{\infty} 2^{-n} \le 2^{-N} \quad (m \ge N),$$

the series converges effectively and uniformly to t on $\mathbf{R}$. Assuming the additional hypotheses for the second part of this exercise, now define a total computable function $H : \mathbf{N} \to \mathbf{N}$ by

$$H(N) \equiv \max\{h(n, N+2) : 0 \le n \le N+2\}.$$

If $N \in \mathbf{N}$, $x, y \in \mathbf{R}$, and $|x - y| \le 2^{-H(N)}$, then

$$
\begin{aligned}
|t(x) - t(y)| \;\le\; & \left| t(x) - \sum_{n=0}^{N+2} 2^{-n} t_n(x) \right| \\
& + \left| \sum_{n=0}^{N+2} 2^{-n} t_n(x) - \sum_{n=0}^{N+2} 2^{-n} t_n(y) \right| \\
& + \left| t(y) - \sum_{n=0}^{N+2} 2^{-n} t_n(y) \right| \\
\le\; & 2^{-N-2} + \sum_{n=0}^{N+2} 2^{-n} |t_n(x) - t_n(y)| + 2^{-N-2} \\
\le\; & 2^{-N-1} + \sum_{n=0}^{N+2} 2^{-n} 2^{-N-2} \\
<\; & 2^{-N}.
\end{aligned}
$$

Hence t is effectively uniformly continuous on $\mathbf{R}$.

It remains to prove that t maps $\mathbf{R}_c$ into $\mathbf{R}_c$. Choose a computable partial function $\theta : \mathbf{N}^2 \to \mathbf{N}$ such that if φ_i is a computable real number generator converging to $x \in \mathbf{R}_c$, then $(n, i) \in \mathrm{domain}(\theta)$ and $\varphi_{\theta(n,i)}$ is a computable real number generator converging to $t_n(x)$. Define a computable partial function $\Psi : \mathbf{N}^2 \to \mathbf{N}$ by

$$
\Psi(i, m) \equiv \sum_{n=0}^{m+1} 2^{-n} \varphi_{\theta(n,i)}(m+1).
$$

Using the s-m-n theorem, construct a total computable function $s : \mathbf{N} \to \mathbf{N}$ such that $\varphi_{s(i)} = \Psi(i, \cdot)$ for each i. Given a computable real number generator φ_i converging to $x \in \mathbf{R}_c$, for each $N \in \mathbf{N}$ we have

$$
\begin{aligned}
|t(x) - \varphi_{s(i)}(N)| \;\le\; & \left| t(x) - \sum_{n=0}^{N+1} 2^{-n} t_n(x) \right| \\
& + \left| \sum_{n=0}^{N+1} 2^{-n} t_n(x) - \sum_{n=0}^{N+1} 2^{-n} \varphi_{\theta(n,i)}(N+1) \right| \\
\le\; & 2^{-N-1} + \sum_{n=0}^{N+1} 2^{-n} |t_n(x) - \varphi_{\theta(n,i)}(N+1)| \\
\le\; & 2^{-N-1} + \sum_{n=0}^{N+1} 2^{-n} 2^{-N-1} \\
<\; & 2^{-N}.
\end{aligned}
$$

Hence $\varphi_{s(i)}$ is a computable real number generator converging to $t(x)$. We now see both that $t(x)$ is a computable real number, and that t is a total computable function from $\mathbf{R}_c$ to $\mathbf{R}_c$.

(4.29.8) Let $f : \mathbf{R}_c \to \mathbf{R}_c$ be the function constructed in the example immediately preceding this set of exercises, and define $g \equiv 1/f$ on $\mathbf{R}_c$. Then g maps $\mathbf{R}_c$ into $\mathbf{R}_c$ and is computable (why?); so, by Theorem (4.27), g is effectively continuous. Suppose that the restriction h of g to $\mathbf{R}_c \cap [0, 1]$ is uniformly continuous. Since each rational number is computable, $\mathbf{R}_c \cap [0, 1]$ is dense in $[0, 1]$; whence, by a standard classical theorem ((3.15.6) of [15]),

h extends to a uniformly continuous function from $[0, 1]$ into $\mathbf{R}$. Another standard theorem ((3.17.10) of [15]) now shows that there exists $M > 0$ such that $|h(x)| \leq M$ for all $x \in [0, 1]$. But this is absurd, since, using the notation used at the end of the discussion of the function f on page 72, we have

$$g(x_N) = 1/f(x_N) \geq 2^N$$

for each N.

Solutions for Chapter 5

(5.3) If

$$T \equiv \{n \in \mathbf{N} : \varphi_n \text{ is total}\}$$

is recursively enumerable, then, by Proposition (5.1), there exists a total computable function $f : \mathbf{N} \to \mathbf{N}$ such that $f \neq g$ for all $g \in T$; this is absurd.

(5.7.1) The mapping $n \mapsto \mathcal{M}_n$ is computable. Also, there is an informally computable mapping which, given a normalised binary Turing machine $\mathcal{M}$ as input, outputs the number of states of $\mathcal{M}$. Composing these two mappings, we obtain **stat**, which is therefore computable.

Now consider $f : \mathbf{N} \to \mathbf{N}$ defined by

$$f(n) \equiv \min\{\mathbf{stat}(k) : \varphi_k = \varphi_n\}.$$

The only normalised binary Turing machine with exactly one state is $(\{0\}, \emptyset, 0, 0)$, which computes the identity mapping **id** on $\mathbf{N}$. Choose an index ν for **id**, and let

$$I \equiv \{n \in \mathbf{N} : \varphi_n = \mathbf{id}\}.$$

Then $\varphi_n = \mathbf{id}$ if and only if $f(n) = 1$. So if f is computable, then I is recursive. This contradicts Theorem (5.5).

(5.7.3) Choose an index ν for the computable partial function $(m, n) \mapsto \varphi_n(m)$ on $\mathbf{N}^2$. Then

$$s(\nu, n) = \min\{i : \varphi_i = \varphi_\nu^{(2)}(\cdot, n)\} = \mathbf{lindex}(n)$$

for each n. Therefore if s is computable, so is **lindex**. This contradicts Exercise (5.7.2).

(5.12) (i) Let

$$S \equiv \{i \in \mathbf{N} : a \in \mathrm{domain}(\varphi_i)\}.$$

The partial function $\Psi : \mathbf{N}^2 \to \mathbf{N}$ defined by

$$\Psi(m, n) \ \ = \ \ 1 \qquad \text{if } n = a \text{ and } m \in K,$$
$$= \ \ \text{undefined} \quad \text{otherwise}$$

is computable (why?). By Corollary (3.9), there exists a total computable function $g : \mathbf{N} \to \mathbf{N}$ such that $\varphi_{g(m)} = \Psi(m, \cdot)$. Then for all $m \in \mathbf{N}$ we have

$$\chi_S \circ g(m) = 1 \ \ \Leftrightarrow \ \ a \in \text{domain}(\varphi_{g(m)})$$
$$\Leftrightarrow \ \ (m, a) \in \text{domain}(\Psi)$$
$$\Leftrightarrow \ \ m \in K;$$

whence $\chi_S \circ g = \chi_K$. It follows that if χ_S is computable, then K is recursive, which contradicts Corollary (4.3).

(5.12) (ii)   Let

$$S \equiv \{i \in \mathbf{N} : \varphi_i \text{ is a constant function}\}.$$

Define a computable partial function $\Psi : \mathbf{N}^2 \to \mathbf{N}$ by

$$\Psi(m, n) \ \ = \ \ 1 \qquad \text{if } m \in K,$$
$$= \ \ \text{undefined} \quad \text{otherwise}.$$

By the s-m-n theorem, there exists a total computable function $g : \mathbf{N} \to \mathbf{N}$ such that $\varphi_{g(m)} = \Psi(m, \cdot)$. For any $m \in \mathbf{N}$ we have

$$\chi_S \circ g(m) = 1 \ \ \Leftrightarrow \ \ \varphi_{g(m)} \text{ is a constant function}$$
$$\Leftrightarrow \ \ \Psi(m, \cdot) \text{ is a constant function}$$
$$\Leftrightarrow \ \ m \in K,$$

so $\chi_S \circ g = \chi_K$. It follows that if χ_S is computable, then K is recursive—a contradiction.

Comment: The technique used in each of the parts of the solution of Exercise (5.12) is known as **reduction** of the given problem to the undecidability of the halting problem.

(5.14.2)   Composing the mapping $n \mapsto \mathcal{T}_n$, defined in Exercise (5.14.1), with the mapping that assigns to each normalised binary Turing machine its index in the enumeration $\mathcal{M}_0, \mathcal{M}_1, \ldots$ we obtain a total computable function $f : \mathbf{N} \to \mathbf{N}$ such that for each $n \in \mathbf{N}$,

$$\varphi_{f(n)}(i) \ \ = \ \ 1 \qquad \text{if } i = n,$$
$$= \ \ \text{undefined} \quad \text{otherwise}.$$

By the Recursion Theorem, there exists $n \in \mathbf{N}$ such that $\varphi_n = \varphi_{f(n)}$. For this n we have

$$i \in \text{domain}(\varphi_n) \quad \Leftrightarrow \quad i \in \text{domain}(\varphi_{f(n)})$$
$$\Leftrightarrow \quad i = n,$$

so $\text{domain}(\varphi_n) = \{n\}$.

Now recall from page 41 that to each computable partial function φ there correspond infinitely many normalised binary Turing machines that compute φ. In particular, computing n as in the last paragraph, we see that there are infinitely many values of $m \neq n$ such that $\varphi_m = \varphi_n$. For any such m,

$$m \notin \{n\} = \text{domain}(\varphi_m),$$

so $\varphi_m(m)$ is not defined, and therefore $m \notin K$. Hence K does not respect indices.

Alternative proof of the first part of (5.14.2).   Define a computable partial function $\Psi : \mathbf{N}^2 \to \mathbf{N}$ by

$$\begin{aligned} \Psi(i, j) \quad &= \quad 1 \qquad\qquad \text{if } j = i, \\ &= \quad \text{undefined} \quad \text{otherwise.} \end{aligned}$$

Using Corollary (3.9), we can find a total computable function $s : \mathbf{N} \to \mathbf{N}$ such that $\varphi_{s(i)} = \Psi(i, \cdot)$ for each i. By the Recursion Theorem, there exists n such that $\varphi_n = \varphi_{s(n)}$. Clearly, $\text{domain}(\varphi_n) = \{n\}$.

(5.14.3) (i)   Define a computable partial function $\Psi : \mathbf{N}^2 \to \mathbf{N}$ by

$$\begin{aligned} \Psi(m, n) \quad &= \quad 0 \qquad\quad \text{if } m = n, \\ &= \quad \varphi_n(n) \quad \text{if } m \neq n. \end{aligned}$$

Using Corollary (3.9), construct a total computable function $s : \mathbf{N} \to \mathbf{N}$ such that $\varphi_{s(m)} = \Psi(m, \cdot)$ for each m. By the Recursion Theorem, there exists i such that $\varphi_i = \varphi_{s(i)}$. Then $\text{domain}(\varphi_i) = K \cup \{i\}$; but $\varphi_i(i) = \Psi(i, i) = 0$, so

$$i \in K = \text{domain}(\varphi_i).$$

(5.14.3) (ii)   This time define the computable partial function $\Psi :$ $\mathbf{N}^2 \to \mathbf{N}$ by

$$\begin{aligned} \Psi(m, n) \quad &= \quad \text{undefined} \quad \text{if } m = n, \\ &= \quad \varphi_n(n) \qquad\quad \text{if } m \neq n. \end{aligned}$$

By Corollary (3.9) and the Recursion Theorem, there exists an index j such that $\varphi_j = \Psi(j, \cdot)$. In this case,

$$\text{domain}(\varphi_j) = \{n \in K : n \neq j\},$$

so $\varphi_j(j)$ is undefined. Hence

$$j \notin K = \text{domain}(\varphi_j).$$

(5.14.4)   Since χ_I is computable, so is $f = j\chi_I + i(1 - \chi_I)$. By the Recursion Theorem, there exists ν such that $\varphi_\nu = \varphi_{f(\nu)}$. If $\nu \in I$, then $\varphi_\nu = \varphi_{f(\nu)} = \varphi_j$; if $\nu \in \mathbf{N} \backslash I$, then $\varphi_\nu = \varphi_{f(\nu)} = \varphi_i$. Thus in either case there exist $m \in I$ and $n \in \mathbf{N} \backslash I$ such that $\varphi_m = \varphi_n$.

Unlike the proof of Rice's Theorem given in the main body of the text, the above proof is fully constructive; that is, it embodies an algorithm for computing m and n such that $m \in I$, $n \in \mathbf{N} \backslash I$, and $\varphi_m = \varphi_n$. (Note that, as you should verify for yourself, our proof of the Recursion Theorem is also fully constructive.)

(5.14.5)   Suppose K is recursive. Then, choosing an index i of the empty partial function ϵ on $\mathbf{N}$, and an index j of the identity function on $\mathbf{N}$, we see that

$$f(n) \; = \; i \quad \text{if } n \in K,$$
$$= \; j \quad \text{if } n \notin K$$

defines a total computable function $f : \mathbf{N} \to \mathbf{N}$. By the Recursion Theorem, there exists ν such $\varphi_{f(\nu)} = \varphi_\nu$. But

$$\nu \in \text{domain}(\varphi_{f(\nu)}) \quad \Rightarrow \quad \varphi_{f(\nu)} \neq \epsilon$$
$$\Rightarrow \quad f(\nu) = j$$
$$\Rightarrow \quad \nu \notin K$$
$$\Rightarrow \quad \nu \notin \text{domain}(\varphi_\nu) = \text{domain}(\varphi_{f(\nu)}),$$

which is absurd.

(5.14.7)   Given a natural number N, choose an index j such that $\varphi_j \neq \varphi_n$ for all $n \leq N$. Define a total computable function $g : \mathbf{N} \to \mathbf{N}$ by

$$g(i) \; = \; j \quad \text{if } i \leq N,$$
$$= \; f(i) \quad \text{if } i > N.$$

Using the Recursion Theorem, we can find i such that $\varphi_i = \varphi_{g(i)}$. By our choice of N, $i > N$ and therefore $g(i) = f(i)$. Thus for each natural number N there exists an index $i > N$ such that $\varphi_i = \varphi_{f(i)}$.

(5.14.8)   Define a computable partial function $\Psi : \mathbf{N} \to \mathbf{N}$ by

$$\Psi(i, j) = \varphi_i(j) + 1 \quad (i, j \in \mathbf{N}).$$

Choose a total computable function $f : \mathbf{N} \to \mathbf{N}$ such that $\varphi_{f(i)} = \Psi(i, \cdot)$ for each i. If φ_i is total, then so is $\varphi_{f(i)}$, and $\varphi_{f(i)}(i) \neq \varphi_i(i)$.

(5.14.9)   Define a computable partial function $\theta : \mathbf{N} \to \mathbf{N}$ by

$$\begin{aligned} \theta(n) &= \varphi_n(n) \quad \text{if } n \in K, \\ &= \text{undefined} \quad \text{otherwise.} \end{aligned}$$

Next, define a computable partial function $\Psi : \mathbf{N}^{n+1} \to \mathbf{N}$ by

$$\Psi(i, u) \equiv \varphi_{\theta(i)}^{(n)}(u) \quad (i \in \mathbf{N}, \ u \in \mathbf{N}^n).$$

By the s-m-n theorem, there exists a total computable function $s : \mathbf{N} \to \mathbf{N}$ such that

$$\varphi_{s(i)}^{(n)} = \Psi(i, \cdot) = \varphi_{\theta(i)}^{(n)}$$

for each i. Another application of the s-m-n theorem now yields a total computable function $g : \mathbf{N} \to \mathbf{N}$ such that $\varphi_{g(i)} = \varphi_i \circ s$ for each i. Let t be the total computable function $s \circ g : \mathbf{N} \to \mathbf{N}$, and consider any k such that φ_k is total. We have

$$\begin{aligned} \varphi_{t(k)}^{(n)} &= \varphi_{s(g(k))}^{(n)} \\ &= \varphi_{\theta(g(k))}^{(n)} \\ &= \varphi_{\varphi_{g(k)}(g(k))}^{(n)} \\ &= \varphi_{\varphi_k(s(g(k)))}^{(n)} \\ &= \varphi_{\varphi_k(t(k))}^{(n)}. \end{aligned}$$

(5.14.10)   As was shown in Exercise (3.11.2), there exist total computable functions $s : \mathbf{N} \to \mathbf{N}$ and $t : \mathbf{N} \to \mathbf{N}$ such that $\varphi_i = \psi_{t(i)}$ and $\psi_i = \varphi_{s(i)}$ for each i. Let f be a total computable function from $\mathbf{N}$ to $\mathbf{N}$. Then $g \equiv s \circ f \circ t$ is a total computable function from $\mathbf{N}$ to $\mathbf{N}$; so, by the Recursion Theorem, there exists m such that $\varphi_m = \varphi_{g(m)}$. Setting $i \equiv t(m)$, we have

$$\psi_i = \psi_{t(m)} = \varphi_m = \varphi_{s(f(t(m)))} = \psi_{f(i)}.$$

(5.14.11) (i)   Writing $\mathbf{0}$ (resp. $\mathbf{1}$) to denote the constant function on $\mathbf{N}$ with each term equal to 0 (respectively 1), for each $k \in \mathbf{N}$ define

$$\psi_{3k} \equiv \mathbf{0}, \ \psi_{3k+1} \equiv \varphi_k, \ \psi_{3k+2} \equiv \mathbf{1}.$$

Choosing indices n_0, n_1 of $\mathbf{0}$ and $\mathbf{1}$, respectively, for each $k \in \mathbf{N}$ define

$$f(3k) = n_0, \ f(3k+1) = k, \ f(3k+2) \equiv n_1, \text{ and } g(k) \equiv 3k+1.$$

Then f and g are total computable functions from $\mathbf{N}$ to $\mathbf{N}$; moreover, $\psi_n = \varphi_{f(n)}$ and $\varphi_n = \psi_{g(n)}$ for each n. Hence, by Exercise (3.11.2), $\psi_0, \psi_1, \psi_2, \ldots$ is an acceptable programming system. Clearly, no three successive terms of this system are equal.

(5.14.11) (ii)   The desired acceptable programming system is given by the sequence

$$\varphi_0, \varphi_1, \varphi_0, \varphi_0, \varphi_1, \varphi_1, \varphi_2, \varphi_0, \varphi_0, \varphi_0, \varphi_1, \varphi_1, \varphi_1, \varphi_2, \varphi_2, \varphi_3, \ldots$$

in which the pattern of indices is

01 00112 000111223 00001111222334 0000011111222233445 ....

Clearly, there exist total computable functions $f, g : \mathbf{N} \to \mathbf{N}$ such that $\psi_n = \varphi_{f(n)}$ and $\varphi_n = \psi_{g(n)}$ for each n; so, by Exercise (3.11.2), the foregoing enumeration of the set of all computable partial functions from $\mathbf{N}$ to $\mathbf{N}$ is an acceptable programming system. (You are invited to provide an exact description of the functions f and g.)

To appreciate the significance of the foregoing examples of acceptable programming systems, first note that, in view of Exercise (5.14.10), the argument used in the example preceding Exercises (5.14), on page 81, applies equally well when the canonical enumeration $\varphi_0, \varphi_1, \ldots$ is replaced by any acceptable programming system. Thus *for each acceptable programming system* $\psi_0, \psi_1, \ldots$, *each computable partial function* ψ, *and each positive integer* k, *there exists* i *such that if* $\psi_i(n)$ *is defined, then so are* $\psi(n)$ *and* $\psi_{i+j}(n)$ $(1 \leq j \leq k)$, *and*

$$\psi_i(n) = \psi_{i+1}(n) = \cdots = \psi_{i+k}(n) = \psi(n).$$

Part (i) of this exercise shows that we cannot drop the words *if $\psi_i(n)$ is defined* from the hypotheses of this last result; in fact, we cannot prove

(A) *If* $\psi_0, \psi_1, \psi_2, \ldots$ *is an acceptable programming system, then there exists* i *such that* $\psi_i = \psi_{i+1} = \psi_{i+2}$.

On the other hand, part (ii) shows that we cannot prove

(B) *If* $\psi_0, \psi_1, \psi_2, \ldots$ *is an acceptable programming system, then there exists* $m > 2$ *such that no m consecutive terms ψ_i are equal.*

Formalising our mathematics within Zermelo-Fraenkel set theory plus the Church-Markov-Turing thesis, we conclude that each of (A) and (B) is independent of the axioms of that formal theory.

(5.17.1)   Given m and n, start $\mathcal{M}_\nu$ on the input n and simultaneously start $\mathcal{M}_m$ on the input m. If $\mathcal{M}_\nu$ halts before $\mathcal{M}_m$ and computes $\psi(n)$, we ignore the continuing computation by $\mathcal{M}_m$ and read the output $\psi(n)$ on the tape of $\mathcal{M}_\nu$. If $\mathcal{M}_m$ halts before $\mathcal{M}_\nu$ and computes $\varphi_m(m)$, we ignore the continuing computation by $\mathcal{M}_\nu$ and apply to the input n a Turing machine that computes φ. If $\mathcal{M}_\nu$ and $\mathcal{M}_m$ complete computations in the same number of steps, we read the output $\psi(n)$ on the tape of $\mathcal{M}_\nu$; there is no ambiguity involved here, since in this case $m \in K$, $n \in \text{domain}(\psi)$, and $\psi(n) = \varphi(n)$.

(5.17.2)   *No.* To see why not, take $\psi(n) \equiv 0$ and $\varphi(n) \equiv 1$ for all $n \in \mathbf{N}$. Suppose there exists a total computable function $f : \mathbf{N} \to \mathbf{N}$ such that for all m,

$$\varphi_{f(m)} \quad = \quad \varphi \quad \text{if } m \in K,$$
$$= \quad \psi \quad \text{if } m \notin K.$$

Then $m \mapsto \varphi_{f(m)}(0)$ is a total computable function on $\mathbf{N}$. Since this function is χ_K, we see that K is recursive, a contradiction.

(5.21.1)   First note that *if a proper subset of $\mathbf{N}$ respects indices and contains indices of ϵ, then it cannot be recursively enumerable*: for if it were recursively enumerable, then, by Proposition (5.18), it would contain the indices of all extensions of ϵ and would therefore equal $\mathbf{N}$. It follows immediately that

$$S \equiv \{i \in \mathbf{N} : \varphi_i = \epsilon\}$$

is not recursively enumerable.

On the other hand, $\mathbf{N} \backslash S$ is recursively enumerable. To see this, for each $i \in \mathbf{N}$ define a total computable function $h : \mathbf{N}^2 \to \mathbf{N}$ by

$$h_i(m, n) \quad = \quad 0 \quad \text{if } \mathcal{M}_i \text{ completes a computation in } m + 1 \text{ steps}$$
$$\text{on the input } n,$$
$$= \quad 1 \quad \text{otherwise.}$$

Let F be a total computable mapping of $\mathbf{N}$ onto $\mathbf{N}^2$, and define a computable partial function $\varphi : \mathbf{N} \to \mathbf{N}$ by

$$\varphi(i) \equiv \min k \, [h_i(F(k)) = 0] .$$

Then $\text{domain}(\varphi) = \mathbf{N} \backslash S$ and so, by Theorem (3.3), $\mathbf{N} \backslash S$ is recursively enumerable.

(5.21.3)   Let S be a proper subset of $\mathbf{N}$ that respects indices, and assume that S is recursive. By Exercise (4.5.3), S and $\mathbf{N} \backslash S$ are proper recursively enumerable subsets of $\mathbf{N}$; clearly, they both respect indices, and one of them contains all indices of the empty function ϵ. The observation at the beginning of the solution of Exercise (5.21.1) shows that this is impossible.

(5.21.5)   Applying the s-m-n theorem, first construct a total computable function $g : \mathbf{N}^2 \to \mathbf{N}$ such that for all m, n, and k,

$$\varphi_{g(m,n)}(k) \quad = \quad \text{undefined} \quad \text{if } \mathcal{M}_m \text{ computes } \varphi_m(m) \text{ in}$$
$$\text{at most } k + 1 \text{ steps,}$$
$$= \quad \varphi_n(k) \qquad \text{otherwise.}$$

Another application of the s-m-n theorem produces a total computable function $h : \mathbf{N}^2 \to \mathbf{N}$ such that $\varphi_{h(m,n)} = \varphi_m(g(\cdot, n))$ for all m and n. Let $f : \mathbf{N} \to \mathbf{N}$ be the total computable function defined by

$$f(m, n) \equiv g(h(m, n), n).$$

Note that

$$f(m, n) \in \operatorname{domain}(\varphi_m) \quad \Leftrightarrow \quad \varphi_m(g(h(m, n), n)) = \varphi_{h(m,n)}(h(m, n))$$
$$\text{is defined}$$
$$\Leftrightarrow \quad h(m, n) \in K.$$

Consider $m \in \mathbf{N}$ such that $I \equiv \operatorname{domain}(\varphi_m)$ respects indices. Given $n \in I$, suppose that $h(m, n) \notin K$. Then our choice of g ensures that

$$\varphi_{f(m,n)} = \varphi_{g(h(m,n),n)} = \varphi_n;$$

so as $n \in I$ and I respects indices, $f(m, n) \in I$. The foregoing now shows that $h(m, n) \in K$—a contradiction. We conclude that $h(m, n)$ must belong to K, and hence that $f(m, n) \in I$. Let $\mathcal{M}_{h(m,n)}$ compute $\varphi_{h(m,n)}(h(m, n))$ in $N + 1$ steps. It follows from the definitions of f and g that

$$\varphi_{f(m,n)}(k) \quad = \quad \varphi_n(k) \qquad \text{if } k \leq N,$$
$$= \quad \text{undefined} \quad \text{otherwise.}$$

Hence $\varphi_{f(m,n)}$ is a finite subfunction of φ_n.

(5.22.1)

$$\begin{array}{ccccc} 11 & 0 & 1111 & 0 & 111111 \\ 1 & & 3 & & 5 \end{array}$$

The encoded set is $\{1, 3, 5\}$.

(5.22.2)   Let $\varphi : \mathbf{N} \to \mathcal{F}$ be given by $\varphi(n) \equiv \{n, n^2\}$. Then $\mu(\varphi(n))$ is the binary number consisting of $(n + 1)$ 1's, followed by 0, followed by $(n^2 + 1)$ 1's; that is,

$$\mu(\varphi(n)) = \sum_{k=0}^{n^2} 2^k + \sum_{k=n^2+2}^{n^2+n+2} 2^k.$$

So $\mu \circ \varphi$ is clearly computable.

(5.24.1)   In the notation of Lemma (5.23), we can decide, for each n, whether $d(n) = \emptyset$ or $d(n) \neq \emptyset$. We can therefore follow the one-one effective enumeration $\psi_0, \psi_1, \ldots$, deleting the unique term with empty domain, to obtain an effective enumeration $\psi_{n_0}, \psi_{n_1}, \ldots$ of the set of nonempty computable partial functions from $\mathbf{N}$ to $\mathbf{N}$. The list $n_0, n_1, \ldots$ is then an effective enumeration of J.

(5.24.2)   First note that we can extract from the proof of Lemma (5.23) an algorithm which, applied to $k \in \mathbf{N}$, computes the code for $d(k)$. Now choose an index ν of θ. Given n, to compute $\psi(n)$ we run $\mathcal{M}_\nu$ on the input n. If $\mathcal{M}_\nu$ completes a computation, we then compute the code c for $d(\theta(n))$ and decide whether or not $d(\theta(n))$ is empty. If $d(\theta(n))$ is nonempty, we decode c and read off the largest element of $d(\theta(n))$.

(5.24.3)   The flaw in the argument stems from the phrase *there exists n such that $S = d(n)$*: although such n must exist, there is no algorithm which, applied to a finite subset X of $\mathbf{N}$, will enable us to compute k such that $X = d(k)$.

(5.25.1)   Define a computable partial function $\Psi : \mathbf{N}^2 \to \mathbf{N}$ by

$$
\begin{aligned}
\Psi(i,j) \;=\; & 1 && \text{if } i \in \text{domain}(\theta), \text{ and} \\
& && \text{either } d(\theta(i)) = \emptyset \text{ and } j = 0, \\
& && \text{or } d(\theta(i)) \neq \emptyset \text{ and } j = 1 + \max \text{domain}(\psi_{\theta(i)}), \\
=\; & \text{undefined} && \text{otherwise.}
\end{aligned}
$$

To confirm that $\Psi(i,j)$ can be computed in the case $d(\theta(i)) \neq \emptyset$, see the solution to Exercise (5.24.2). Choose a total computable function $s : \mathbf{N} \to \mathbf{N}$ such that $\varphi_{s(i)} = \Psi(i, \cdot)$ for each i. By the Recursion Theorem, there exists n such that $\varphi_n = \varphi_{s(n)}$. Clearly, domain$(\varphi_n)$ is disjoint from domain$(\psi_{\theta(n)})$. Since domain(φ_n) is finite, $\theta(n)$ is defined, so domain(φ_n) is nonempty.

Now suppose that there exists a computable partial function $\gamma : \mathbf{N} \to \mathbf{N}$ with properties (i) and (ii) of the statement of this exercise. The foregoing ensures that there exists $n \in F$ such that domain(φ_n) is nonempty and disjoint from domain$(\psi_{\gamma(n)})$. This contradicts property (ii) of γ.

(5.25.2)   Define a computable partial function $\Psi : \mathbf{N}^2 \to \mathbf{N}$ by

$$
\begin{aligned}
\Psi(i,j) \;=\; & 1 && \text{if } i \in \text{domain}(\theta) \text{ and } j = \theta(i) + 1, \\
=\; & \text{undefined} && \text{otherwise.}
\end{aligned}
$$

By the s-m-n theorem, there exists a total computable function $s : \mathbf{N} \to \mathbf{N}$ such that $\varphi_{s(i)} = \Psi(i, \cdot)$ for each i. Applying the Recursion Theorem to compute n such that $\varphi_n = \varphi_{s(n)}$, we see that $n \in F$ and that domain$(\varphi_n) = \{\theta(n) + 1\}$.

Now let $\gamma : \mathbf{N} \to \mathbf{N}$ be a computable partial function with the properties (i) and (ii) described in Exercise (5.25.1). Define a computable partial function $\theta : \mathbf{N} \to \mathbf{N}$ by

$$
\begin{aligned}
\theta(n) &= \max \operatorname{domain}(\psi_{\gamma(n)}) && \text{if } n \in \operatorname{domain}(\gamma) \text{ and } d(\gamma(n)) \neq \emptyset, \\
&= 0 && \text{if } n \in \operatorname{domain}(\gamma) \text{ and } d(\gamma(n)) = \emptyset, \\
&= \text{undefined} && \text{otherwise.}
\end{aligned}
$$

Then $\operatorname{domain}(\theta) \supset \operatorname{domain}(\gamma) \supset F$. By the first part of this exercise, there exists $n \in F$ such that $\operatorname{domain}(\varphi_n) = \{\theta(n) + 1\}$. If $d(\gamma(n)) = \emptyset$, then φ_n has domain $\{1\}$; if $d(\gamma(n)) \neq \emptyset$, then $\operatorname{domain}(\varphi_n)$ contains a single element, which is greater than each element of $d(\gamma(n))$. Hence in either case, $\varphi_n \neq \psi_{\gamma(n)}$—a contradiction.

(5.29.2)   By Exercise (5.24.1),

$$
J \equiv \{j \in \mathbf{N} : \operatorname{domain}(\psi_j) \neq \emptyset\}
$$

is recursively enumerable. Clearly, $i \in \mathbf{N}\backslash S$ if and only if there exists $j \in J$ such that $\psi_j \subset \varphi_i$. Hence, by Theorem (5.28), $\mathbf{N}\backslash S$ is recursively enumerable.

(5.29.3)   Define a computable partial function $\Psi : \mathbf{N}^2 \to \mathbf{N}$ by

$$
\begin{aligned}
\Psi(i,j) &= 1 && \text{if } i \in \operatorname{domain}(\theta), \ d(\theta(i)) = \emptyset, \text{ and } j = 0, \\
&= \psi_{\theta(i)}(j) + 1 && \text{if } i \in \operatorname{domain}(\theta) \text{ and } j \in d(\theta(i)), \\
&= \text{undefined} && \text{otherwise.}
\end{aligned}
$$

Choose a total computable function $s : \mathbf{N} \to \mathbf{N}$ such that $\varphi_{s(i)} = \Psi(i, \cdot)$ for each i. By the Recursion Theorem, there exists n such that $\varphi_n = \varphi_{s(n)}$. Clearly, $\operatorname{domain}(\varphi_n) = \operatorname{domain}(\Psi(n, \cdot))$ is finite; so $\theta(n)$ is defined, and therefore $\operatorname{domain}(\varphi_n)$ is nonempty. Also, if $j \in \operatorname{domain}(\psi_{\theta(n)})$, then $\varphi_n(j)$ is defined and $\varphi_n(j) = \psi_{\theta(n)}(j) + 1$; whence $\psi_{\theta(n)} \not\subset \varphi_n$.

(5.29.4)   The set
$$
I \equiv \{i \in \mathbf{N} : \varphi_i \neq \epsilon\}
$$
is recursively enumerable, by Exercise (5.29.2), and respects indices. Suppose there exists a total computable function s with the stated properties relative to I. Then, by Exercise (5.29.3), there exists n such that $\operatorname{domain}(\varphi_n)$ is nonempty and finite, and such that either $\psi_{s(n)} = \epsilon$ or $\psi_{s(n)} \not\subset \varphi_n$. Since $n \in I$, there exists $k \in I$ such that $\psi_{s(n)} = \varphi_k$; so $\psi_{s(n)} \neq \epsilon$, and therefore $\psi_{s(n)} \not\subset \varphi_n$. This contradicts the assumed property (ii) of s.

Solutions for Chapter 6

(6.1.1) (i)   Take $\gamma_i \equiv \varphi_i$ for each i. Then B1 is automatically satisfied. But if B2 holds, then $\{i \in \mathbf{N} : \varphi_i(0) = 0\}$ is a recursive set; since this set clearly respects indices, is nonempty, and is a proper subset of $\mathbf{N}$, this contradicts Rice's Theorem.

(ii)   Take $\gamma_i(n) \equiv 0$ for all i and n.

(6.1.2)   It is clear that B1 is satisfied. On the other hand, the function **costs'** $: \mathbf{N}^3 \to \mathbf{N}$, defined by

$$
\begin{aligned}
\mathbf{costs'}(i, n, k) \quad &= \quad \mathbf{costs}(i, n, k) \quad \text{if } i \neq j, \\
&= \quad 1 \qquad\qquad\qquad \text{if } i = j \text{ and } k = 0, \\
&= \quad 0 \qquad\qquad\qquad \text{if } i = j \text{ and } k \geq 1,
\end{aligned}
$$

is computable, and

$$
\begin{aligned}
\mathbf{costs'}(i, n, k) \quad &= \quad 1 \quad \text{if } \gamma_i'(n) = k, \\
&= \quad 0 \quad \text{otherwise.}
\end{aligned}
$$

So Γ' satisfies B2.

Using Γ' as our complexity measure, we see that the cost of computing $\varphi_j(n)$ is 0; in other words, it costs nothing to decide whether or not n belongs to the recursive set S. In the particular case where S is taken as the set of all prime numbers, this situation certainly does not reflect reality: it is well known that testing integers for primality is an extremely costly business. Indeed, all known algorithms for primality testing have cost that grows exponentially as a function of the size, in bits, of the integer under test. For further information on this topic, see Chapter 4 of [33].

(6.1.3)   It is clear from axiom B1, applied to φ_i and γ_i, that

$$
\mathrm{domain}(\gamma_i') = \mathrm{domain}(\varphi_i).
$$

On the other hand, given positive integers n and k, and using axiom B2, we can decide whether or not there exists $j \leq k$ such that $\gamma_i(n) = j$. If such j exists, then, by B1, $\varphi_i(n)$ is defined, so $\gamma_i'(n)$ is defined; moreover, by comparing $f(\varphi_i(n))$ with $k - j$ we can decide whether or not $\gamma_i'(n)$ equals k. If, however, no such j exists, then it is impossible for $\gamma_i'(n)$ to equal k. Thus the function **costs'** $: \mathbf{N} \to \mathbf{N}$, defined by

$$
\begin{aligned}
\mathbf{costs'}(i, n, k) \quad &= \quad 1 \quad \text{if } \gamma_i'(n) = k, \\
&= \quad 0 \quad \text{otherwise,}
\end{aligned}
$$

is computable.

(6.1.4)   We have

$$G(i, n, k) = \sum_{j=0}^{t(k)} \textbf{costs}(i, n, j).$$

Since t and **costs** are total computable functions, so is G.

(6.1.5)   The existence of s is a simple consequence of the *s-m-n* theorem. We compute $G(n, i, j, k)$ as follows. Compute first $v(n)$ and then **costs**$(i, v(n), j)$. If the latter equals 1, then $\gamma_i(v(n)) = j$, $\varphi_i(v(n))$ is defined, and we can compute **costs**$(\varphi_i(v(n)), n, k)$; if that equals 1, then $\varphi_{s(i)}(n) = \varphi_{\varphi_i \circ v(n)}(n)$ is defined and we set $G(n, i, j, k) \equiv \gamma_{s(i)}(n)$. On the other hand, if

  either **costs**$(i, v(n), j) = 0$
  or **costs**$(i, v(n), j) = 1$ and **costs**$(\varphi_i(v(n)), n, k) = 0,$

we set $G(n, i, j, k) \equiv 0$.

(6.3)   Since

$$\Phi(i, n) = \min k[\textbf{costs}(i, n, k) = 1],$$

the computability of Φ follows from Exercise (2.7.3).

(6.5.1)   Define

$$\gamma_i' \equiv 1 + \gamma_i + \varphi_i;$$

then, by Exercise (6.1.3), $\Gamma' \equiv \gamma_0', \gamma_1', \gamma_2', \ldots$ is a complexity measure. Let $F : \mathbf{N}^2 \to \mathbf{N}$ be a total computable function. By the *s-m-n* theorem, there exists a total computable function $s : \mathbf{N} \to \mathbf{N}$ such that

$$\varphi_{s(i)}(n) = F(n, \gamma_i(n))$$

for each i and for all $n \in \text{domain}(\gamma_i)$. Applying the Recursion Theorem, we obtain an index ν such that $\varphi_{s(\nu)} = \varphi_\nu$. Thus

$$\gamma_\nu' = 1 + \gamma_\nu + \varphi_{s(\nu)} = 1 + \gamma_\nu + F(\cdot, \gamma_\nu(\cdot)),$$

so $\gamma_\nu'(n) > F(n, \gamma_\nu(n))$ for all $n \in \text{domain}(\varphi_\nu)$.
  Despite appearances, this result does not contradict Theorem (6.4), since it does not guarantee that $\gamma_\nu'(n) > F(n, \gamma_\nu(n))$ infinitely often. Indeed, it follows from Theorem (6.4) that domain(φ_ν) must be finite.

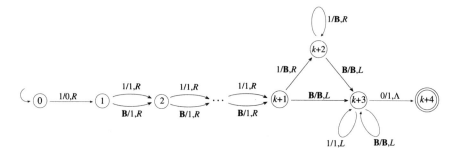

FIGURE 29. The Turing machine $\mathcal{T}_k$ in solution (6.7.1).

(6.7.1) Let

$$\bar{\gamma}_i(n) \quad \equiv \quad \text{the number of distinct cells visited by } \mathcal{M}_i$$
$$\text{during the computation of } \varphi_i(n).$$

For each k let $\mathcal{T}_k$ be the Turing machine described in Figure 29. Then $\mathcal{T}_k$ computs the constant mapping $n \mapsto k$ on $\mathbf{N}$. There is a one-one total computable function $h : \mathbf{N} \to \mathbf{N}$ such that $\mathcal{T}_k = \mathcal{M}_{h(k)}$ for each k. Moreover, as is easily verified, the range of h is a recursive subset of $\mathbf{N}$, and the partial function $\varphi : \mathbf{N} \to \mathbf{N}$ defined by

$$\varphi(i) \quad = \quad h^{-1}(i) \qquad \text{if } i \in \text{range}(h),$$
$$= \quad \text{undefined} \quad \text{if } i \notin \text{range}(h)$$

is computable (cf. Exercise (3.5.1)). Now define

$$\gamma_i(n) \quad = \quad \max\{0, \bar{\gamma}_i(n) - \varphi(i) - 2\} \quad \text{if } i \in \text{range}(h),$$
$$= \quad \bar{\gamma}_i(n) \qquad\qquad\qquad \text{if } i \notin \text{range}(h).$$

It is straightforward to verify that $\Gamma \equiv \gamma_0, \gamma_1, \ldots$ is a complexity measure. Given a total computable function $F : \mathbf{N}^2 \to \mathbf{N}$, choose $k > F(0,0)$ and set $i \equiv h(k)$. Then $\bar{\gamma}_i(0) = k + 2$, so that

$$\gamma_i(0) = \max\{0, k + 2 - \varphi(h(k)) - 2\} = 0,$$

and therefore

$$\varphi_i(0) = k > F(0,0) = F(0, \gamma_i(0)).$$

(6.9) The total computable function $v : \mathbf{N} \to \mathbf{N}$, with values $v(n)$ equal to

$$0, 0, 1, 0, 1, 2, 0, 1, 2, 3, 0, 1, 2, 3, 4, \ldots$$

for $n = 0, 1, 2, \ldots$, has the property that for each $k \in \mathbf{N}$ there are infinitely many values of n with $v(n) = k$.[3]

Given a total computable function $t : \mathbf{N} \to \mathbf{N}$, and defining the total function $f : \mathbf{N} \to \mathbf{N}$ as in the statement of this exercise, we see from B1 and Exercise (6.1.4) that f is computable. For each n such that $v(n) \in \mathbf{IND}(f)$, $\varphi_{v(n)} = f$ is total, as is therefore $\gamma_{v(n)}$. Since $\varphi_{v(n)}(n) \neq \varphi_{v(n)}(n) + 1$, we have $\gamma_{v(n)}(n) > t(n)$.

(6.11.1)   Fix the natural number n. For each element u of $\{0,1\}^{n+1}$ define a Turing machine module $\mathcal{T}(u)$ as follows. If (the unary representation of) an input $k \in \mathbf{N}$ is written in the leftmost cells of the tape, and if $\mathcal{T}(u)$ is in its start state, with the read/write head against the leftmost cell, $\mathcal{T}(u)$ first compares k with n. If $k \leq n$, then $\mathcal{T}(u)$ writes 1 in the leftmost cell if $P_k^{n+1}(u) = 0$, 11 in the two leftmost cells if $P_k^{n+1}(u) = 1$; leaves all other cells blank; and parks the read/write head. If $k > n$, then $\mathcal{T}(u)$ writes (the unary representation of) k in the leftmost cells, leaves all other cells blank, and enters a special state q_u which is not its halt state or its start state, and from which there are no transitions. It is easy to see that the construction of $\mathcal{T}(u)$ can be carried out so that the number of steps it requires to complete its computation on the input $k \leq n$ is bounded by $b(k)$ for some total computable function $b : \mathbf{N} \to \mathbf{N}$ that depends on n but is independent of k and u.

Given a total computable function $f : \mathbf{N} \to \{0,1\}$ and an index ν of f, let

$$u \equiv (f(0), \ldots, f(n)) \in \{0,1\}^{n+1}.$$

Renaming the states of $\mathcal{M}_\nu$, we may assume that q_u is the start state of $\mathcal{M}_\nu$, and that $\mathcal{T}(u)$ and $\mathcal{M}_\nu$ have no other state in common. We then append $\mathcal{M}_\nu$ to $\mathcal{T}(u)$, and add transitions that ensure that if the resulting Turing machine enters the halt state of $\mathcal{M}_\nu$, it does not halt immediately but, without moving the read/write head, passes to the halt state of $\mathcal{T}(u)$ and then halts. Finally, we rename the states of this Turing machine to obtain a normalised binary Turing machine $\mathcal{M}_i$ such that $i \in \mathbf{IND}(f)$ and such that $\gamma_i(k) \leq b(k)$ for $k = 0, \ldots, n$.

(6.12)   For all total computable functions f, g from $\mathbf{N}$ to $\mathbf{N}$ we have

$$(\gamma_i(n) \leq f(n) \text{ and } \gamma_i(n) \leq g(n)) \Leftrightarrow \gamma_i(n) \leq \min\{f(n), g(n)\};$$

whence

$$C_f \cap C_g = C_{\min\{f,g\}}.$$

---

[3]Calude, in a private communication, has suggested that the mapping ν be known as the *Halmos sequence*, since it is the sequence in which Halmos recommends the writing of chapters in a book. (See the latter's article in *How to Write Mathematics*, American Mathematical Society, Providence R.I., 1973.)

Now define

$$f(n) \quad = \quad 1 \quad \text{if } n \text{ is even,}$$
$$= \quad 0 \quad \text{if } n \text{ is odd,}$$

and set $g \equiv 1 - f$. Given a complexity measure Γ, construct a new complexity measure Γ' as follows: choosing $i \in \mathbf{IND}(f)$, $j \in \mathbf{IND}(g)$, and an index k for the constant function $\mathbf{1}$, define

$$\gamma'_n \quad = \quad \gamma_n \quad \text{if } n \notin \{i, j, k\},$$
$$= \quad f \quad \text{if } n = i,$$
$$= \quad g \quad \text{if } n = j,$$
$$= \quad 1 \quad \text{if } n = k.$$

(You should verify that Γ' is a complexity measure.) For each total computable function $h : \mathbf{N} \to \mathbf{N}$ let C'_h denote the complexity class of h relative to Γ'. Suppose $C'_f \cup C'_g = C'_h$. Then $f \in C'_h$, so $h(n) \geq 1$ for all sufficiently large even n; and $g \in C'_h$, so $h(n) \geq 1$ for all sufficiently large odd n. Hence $h(n) \geq 1$ almost everywhere, so that $\varphi_k \in C'_h$, and therefore either $\varphi_k \in C'_f$ or else $\varphi_k \in C'_g$. This is plainly absurd. Thus there does not exist a total computable function $h : \mathbf{N} \to \mathbf{N}$ such that $C'_f \cup C'_g = C'_h$.

(6.15.1)   The intervals $[k_i, k_{i+2})$ $(i = 0, 2, 4, \ldots, 2n)$ are disjoint, and there are $n + 1$ of them. So there exists at least one i such that $[k_i, k_{i+2})$ does not contain any point c_j. For this i and all j $(1 \leq j \leq n)$, we have $c_j \notin [k_i, k_{i+2})$, so $c_j \notin [k_i, k_{i+1}]$.

(6.15.3)   For convenience write

$$P(m, k, n) \quad \Leftrightarrow \quad \varphi_m(n) \leq k \text{ and}$$
$$\forall i < n \; (\gamma_i(n) < k \text{ or } \neg(\gamma_i(n) \leq F(n, k))).$$

Define a computable partial function function $\chi : \mathbf{N}^3 \to \mathbf{N}$ by

$$\chi(m, k, n) \quad = \quad 0 \qquad\qquad\quad \text{if } P(m, k, n),$$
$$= \quad 1 \qquad\qquad\quad \text{if } n \in \text{domain}(\varphi_m) \text{ and } \neg P(m, k, n),$$
$$= \quad \text{undefined} \quad \text{otherwise.}$$

We prove that if φ_m is total, then for each n,

$$\mathcal{D}(m, n) \equiv \{k \in \mathbf{N} : \chi(m, k, n) = 0\}$$

is nonempty. To this end, define a computable partial function $\kappa : \mathbf{N}^3 \to \mathbf{N}$ by

$$\kappa(m, j, n) \quad = \quad \varphi_m(n) \qquad\qquad\quad \text{if } j = 0,$$
$$= \quad F(n, \kappa(m, j - 1, n)) \quad \text{if } j \geq 1.$$

If φ_m is total, then

$$\varphi_m(n) = \kappa(m, 0, n) < \kappa(m, 1, n) < \cdots.$$

Since there are at most n values $\gamma_i(n)$ with $0 \le i < n$, Exercise (6.15.1) shows that there exists r $(0 \le r \le 2n + 1)$ such that for all $i < n$,

$$\gamma_i(n) \notin [\kappa(m, r, n), \kappa(m, r + 1, n)]$$

and therefore either $\gamma_i(n) < \kappa(m, r, n)$ or

$$\neg(\gamma_i(n) \le \kappa(m, r + 1, n) = F(n, \kappa(m, r, n))).$$

Hence $P(m, \kappa(m, r, n), n)$ holds, and therefore $\kappa(m, r, n) \in \mathcal{D}(m, n)$.

We now see from Exercise (2.7.3) that

$$\Psi(m, n) \equiv \min k\, [\chi(m, k, n) = 0]$$

defines a computable partial function $\Psi : \mathbf{N}^2 \to \mathbf{N}$. By the s-m-n theorem, there exists a total computable function $s : \mathbf{N} \to \mathbf{N}$ such that $\varphi_{s(m)} = \Psi(m, \cdot)$. If φ_m is total, then, by the foregoing, $\varphi_{s(m)}$ is total, and $P(m, \varphi_{s(m)}(n), n)$ holds for each n; whence $\varphi_m(n) \le \varphi_{s(m)}(n)$ and

$$\forall i < n\ (\gamma_i(n) < \varphi_{s(m)}(n) \text{ or } \neg(\gamma_i(n) \le F(n, \varphi_{s(m)}(n)))).$$

So if $\varphi_{s(m)}(n) \le \gamma_i(n) \le F(n, \varphi_{s(m)}(n))$, then $n \le i$.

(6.16) Using Theorem (6.4), construct a total computable function $F : \mathbf{N}^2 \to \mathbf{N}$ such that $\gamma_i(n) \le F(n, \gamma_i'(n))$ almost everywhere. We may assume that

$$n < F(m, n) < F(m, n + 1)$$

for all m and n. According to Theorem (6.13), for each total computable function $t : \mathbf{N} \to \mathbf{N}$ there exists a total computable function $f : \mathbf{N} \to \mathbf{N}$ such that $f(n) \ge t(n)$ for all n, and such that if $f(n) \le \gamma_i(n) \le F(n, f(n))$, then $n \le i$. Consider any i, n such that $n > i$, $\gamma_i'(n) \le f(n)$, and $\gamma_i(n) \le F(n, \gamma_i'(n))$. If $f(n) \le \gamma_i(n)$, then

$$f(n) \le \gamma_i(n) \le F(n, \gamma_i'(n)) \le F(n, f(n)),$$

so $n \le i$, a contradiction; hence $\gamma_i(n) < f(n)$. It follows that $C_{f'} \subset C_f$. The reverse inequality follows from the hypothesis that $\gamma_i'(n) \le \gamma_i(n)$ for all i and all $n \in \text{domain}(\varphi_i)$.

(6.19) If $\varphi_i(n)$ is defined, then, by B1, so is $\gamma_i(n)$. So, using Exercise (6.1.4), we can decide, for each $j < n$, whether or not $P(i, j, n)$ holds. There will be at most n values of $j < n$ for which $P(i, j, n)$ holds, and therefore at most n corresponding values $\varphi_j(n)$. Straightfoward computations enable us to find k, from among the $n + 1$ values $0, 1, \ldots, n$, such that $k \ne \varphi_j(n)$ for all $j < n$ for which $P(i, j, n)$ holds. (Note that in view of B1, $\varphi_j(n)$ is defined for each such j.) Hence $\Psi(i, n)$ is defined and at most n.

(6.22.1) If $n \leq i$, then $\mathcal{C}(e, i, n)$ is defined to be $\emptyset$, which is certainly both finite and recursive. In particular, $\mathcal{C}(e, i, 0)$ is defined, finite, and recursive. Assume that for $0 \leq m < n$, if $\mathcal{C}(e, i, m)$ is defined, then it is finite and recursive. If $\mathcal{C}(e, i, n)$ is defined, then in order to complete our inductive proof we need only deal with the case $i < n$. Then $\mathcal{C}(e, i, m)$ is defined— and therefore both finite and recursive—for $0 \leq m < n$, and $\gamma_{s(e,j+1)}(n)$ is defined whenever $i \leq j < n$. Given $j \in \mathbf{N}$, we can decide whether or not $i \leq j < n$. Moreover, using our induction hypothesis and Exercise (6.1.4), we can decide, for each j with $i \leq j < n$, whether or not

$$j \notin \bigcup_{m=0}^{n-1} \mathcal{C}(e, i, m) \text{ and } \gamma_j(n) < F(n, \gamma_{s(e,j+1)}(n));$$

so $\mathcal{C}(e, i, n)$, which is obviously finite, is recursive.

(6.24.1) Let f be the identity function $\mathbf{id} : \mathbf{N} \to \mathbf{N}$. Then f is computed by the normalised binary Turing machine $\mathcal{M} \equiv \{\{0\}, \emptyset, 0, 0\}$ (cf. the solution to Exercise (5.7.1)). Let ν be the index of $\mathcal{M}$, and define a complexity measure $\Gamma \equiv \gamma_0, \gamma_1, \ldots$ by

$$\gamma_i(n) \equiv \gamma_i^*(n) + |\nu - i|,$$

where γ_i^* is defined as in the proof of the Speed-up Theorem. Consider any total computable function $F : \mathbf{N}^2 \to \mathbf{N}$ such that $F(m, n+1) \geq F(m, n)$ for all m, n. For each $j \in \mathbf{IND}(f)$ with $j \neq \nu$, and for all $n \in \mathbf{N}$, we have

$$F(n, \gamma_j(n)) \geq \gamma_j^*(n) + |\nu - j| > \gamma_j^*(n) \geq 1 = \gamma_\nu(n).$$

Hence f is not F-speedable relative to the complexity measure Γ.

(6.24.3) Let Γ be any complexity measure, and take $F(m, n) \equiv n + 1$ for all $m, n \in \mathbf{N}$. Then F-speedable functions exist, by the Speed-up Theorem. Let f be any one of them. Since

$$\{\gamma_i(0) : i \in \mathbf{IND}(f)\}$$

is a set of nonnegative integers, it has a least member; that is, there exists $\nu \in \mathbf{IND}(f)$ such that

$$\gamma_\nu(0) = \min\{\gamma_i(0) : i \in \mathbf{IND}(f)\}.$$

For all $j \in \mathbf{IND}(f)$ we have

$$F(0, \gamma_j(0)) = \gamma_j(0) + 1 > \gamma_\nu(0).$$

References

[1] Aberth, Oliver: *Computable Analysis*. New York: McGraw-Hill 1980.

[2] Barendregt, H.P.: *The Lambda Calculus: its Syntax and Semantics*. Amsterdam: North-Holland 1981.

[3] Barwise, J. (ed.): *Handbook of Mathematical Logic*. Amsterdam: North-Holland 1977.

[4] Beeson, Michael J.: *Foundations of Constructive Mathematics*. New York-Heidelberg-Berlin: Springer-Verlag 1985.

[5] Bishop, Errett, and Bridges, Douglas S.: *Constructive Analysis* (Grundlehren der math. Wissenschaften 279). New York-Heidelberg-Berlin: Springer-Verlag 1985.

[6] Blum, M.: A machine-independent theory of the complexity of recursive functions. J. Assoc. Comput. Mach. 14, 322-336 (1967).

[7] Blum, M.: On effective procedures for speeding-up algorithms. ACM Symposium on Theory of Computing, 43-53 (1969).

[8] Bridges, D.S., and Richman, Fred: *Varieties of Constructive Mathematics* (London Mathematical Society Lecture Notes 97). Cambridge: Cambridge University Press 1987.

[9] Calude, C.: *Theories of Computational Complexity*. Amsterdam: North-Holland 1988.

[10] Calude, C., and Zimand, M.: Recursive Baire classification and speedable functions, Zeitschr. math. Logik Grundlagen Math. 38, 169-178 (1992).

[11] Čeitin, G.S.: Algorithmic operators in constructive complete separable metric spaces (Russian). Doklady Akad. Nauk 128, 49-52 (1959).

[12] Chaitin, G.: Lisp Program-size Complexity. Appl. Math. and Comput. 49, 79-93 (1992).

[13] Cohen, Paul J.: *Set Theory and the Continuum Hypothesis*. New York: Benjamin 1966.

[14] Cutland, N.J.: *Computability, an Introduction to Recursive Function Theory*. Cambridge: Cambridge University Press 1980.

[15] Dieudonné, J.: *Foundations of Modern Analysis*. New York: Academic Press 1960.

[16] Dowling, W.F.: Computer Viruses: diagonalization and fixed points. Notices Amer. Math. Soc. 37(7), 858-861 (1990).

[17] Gödel, Kurt: *The Consistency of the Continuum Hypothesis* (Ann. Math. Studies 3). Princeton: Princeton University Press 1940.

[18] Halmos, P.R.: *Naive Set Theory*. New York-Heidelberg-Berlin: Springer-Verlag 1974.

[19] Kalmár, L.: An argument against the plausibility of Church's thesis. In: Heyting, A. (ed.), *Constructivity in Mathematics* (Proceedings of the Colloquium at Amsterdam, 1957). Amsterdam: North-Holland 1959.

[20] Kfoury, A.J., Moll, R.N., and Arbib, M.A.: *A Programming Approach to Computability*. New York-Heidelberg-Berlin: Springer-Verlag 1982.

[21] Ko, Ker-I: *Complexity Theory of Real Functions*. Boston-Basel-Berlin: Birkhaüser 1991.

[22] Kreisel, G., Lacombe, D., and Schoenfield, J.: Partial recursive functions and effective operations. In: Heyting, A. (ed.), *Constructivity in Mathematics* (Proceedings of the Colloquium at Amsterdam, 1957). Amsterdam: North-Holland 1959.

[23] Machtey, M., and Young, P.: *An Introduction to the General Theory of Algorithms*. Amsterdam: North-Holland 1978.

[24] Odifreddi, P.: *Classical Recursion Theory, Volume* 1. Amsterdam: North-Holland 1990.

[25] Péter, Rózsa: *Recursive Functions in Computer Theory*. Chichester: Ellis-Horwood 1981.

[26] Pour-El, Marian B., and Richards, Jonathan I.: *Computability in Analysis and Physics*. New York-Heidelberg-Berlin: Springer-Verlag 1988.

[27] Rayward-Smith, V.J.: *A First Course in Computability*. Oxford: Blackwell Scientific Publications 1986.

[28] Rogers, Hartley Jr.: *Theory of Recursive Functions and Effective Computability*. New York: McGraw-Hill 1967.

[29] Salomaa, Arto: *Computation and Automata*. Cambridge: Cambridge University Press 1985.

[30] Schnorr, C.P.: Does the computational speed-up concern programming? In Nivat, M. (ed.): *Automata, Languages and Programming*. Amsterdam: North-Holland 1973.

[31] Todd, John: *Introduction to the Constructive Theory of Functions*. Basel: Birkhaüser Verlag 1963.

[32] Weihrauch, Klaus: *Computability*. New York-Heidelberg-Berlin: Springer-Verlag 1987.

[33] Wilf, H.S.: *Algorithms and Complexity*. Englewood Cliffs, New Jersey: Prentice-Hall Inc. 1986.

[34] Wood, Derick: *Theory of Computation*. New York: Harper & Row 1987.

If you are interested in pursuing the study of computability, try reading [28], [29], or [24]. The first of these is rather dated, but certainly a classic; the other two are likely to become classics. Other references for computability are [14], [20], [23], [27], [32], and [34]. A comprehensive account of abstract complexity theories, including several not mentioned in this book, is given in [9]. For an authoritative account of the complexity-theoretic analysis of real variables see [21]. Accounts of recursive and constructive mathematics using informal intuitionistic logic are found in [4], [5], and [8].

Index

Graduate Texts in Mathematics

continued from page ii